T0180840

Modeling and Control of Power Electronic Converters for Microgrid Applications

Yang Han

Modeling and Control of Power Electronic Converters for Microgrid Applications

 Springer

Yang Han
University of Electronic Science and Technology of China
Chengdu, China

ISBN 978-3-030-74515-8   ISBN 978-3-030-74513-4   (eBook)
https://doi.org/10.1007/978-3-030-74513-4

This Springer imprint is published by the registered company Springer Nature Switzerland AG
The registered company address is: Gewerbestrasse 11, 6330 Cham, Switzerland

Preface

Distributed generation (DG) using renewable energy resources (RES) such as wind turbines and solar power plants has been widely used in recent years. However, the increasing penetration of the DG systems may bring problems like inverse power flow, voltage deviation, and voltage fluctuation to the distribution networks. To achieve better operation of multiple DG units, the microgrid concept using the coordinated control strategies among parallel-connected DG interfacing converters has received lots of attention both in academia and industry. Compared to the conventional distribution system, the microgrid can operate in both grid-connected and autonomous islanding modes, offering more reliable power to the critical loads. In the islanded mode, each DG unit should be able to supply a certain amount of the total load proportional to its power rating.

To achieve the power-sharing requirement with only local measurement and eliminating external high-bandwidth communication links among the DG units, the frequency and voltage droop control methods are widely adopted and the "plug-and-play" functions can be applied to enhance the reliability of the system. However, the conventional droop control method has several drawbacks including a trade-off between power-sharing accuracy and voltage deviation, and a high dependency on the line impedances. Although accurate real power sharing can be achieved by the frequency droop method, the poor reactive power sharing due to impedance mis-match of the DG feeders and the different ratings of the DG units is inevitable. Hence, the modeling and control of power electronic converters are crucial to ensure a stable and reliable operation of the microgrid system.

In the past 10 years, the author was involved in lots of academic and industrial projects related to power electronics, microgrids, power quality monitoring, and compensation for distribution systems. Meanwhile, the author was responsible for teaching automatic control theory, power electronics, and renewable power systems to undergraduate students as well as the master's course on simulation software to electrical engineering students. The experience from the practical projects was successfully applied for classroom teaching and academic research; several research papers were published in *IEEE Transactions* and presented at various IEEE or domestic conferences, which provides sufficient support for this book.

In retrospect, the author was invited to give a panel presentation at the 24th International Conference on Electrical Engineering (ICEE 2018) in Seoul, Korea, and various invited talks or tutorials in other conferences. The idea behind this book is inspired by the invitation e-mail from the editorial team at Springer in 2019 for the possible book proposal, after the invited tutorial at 2019 IEEE PES ISGT conference on "modeling and control of voltage source converter for power quality condition and smart microgrid." Hence, the author aims to collect some of the previous work and update the presentation, and lots of new analysis, case studies, benchmark system from practical experience and course teaching are added to enrich the content of the book. This book was expected to be finalized by early 2020. However, due to the pandemic, the progress was slow. Fortunately, the manuscript got finalized in early 2021. It is expected that this book might be helpful for readers both in academia and industry.

The author would like to express his sincere gratitude to the National Natural Foundation of China, Natural Science Foundation of Guangdong Province, and Sichuan Province Key Research and Development Project for their financial support, and also China Scholarship Council (CSC) for the financial support during his academic visit to the Department of Energy Technology, Aalborg University, Denmark, from March 2014 to March 2015. Special gratitude is conveyed to Prof Josep M. Guerrero; Prof Frede Blaabjerg; Prof Zhe Chen from Aalborg University, Denmark; Prof Ernane A. A. Coelho from Universidade Federal de Uberlandia, Brazil; and Prof Chen Chen, from Shanghai Jiaotong University, China, for their kind help, guidance, and cooperation in the past years. In addition, the author would also like to express gratitude to his master's students, Pan Shen, Hong Li, Xu Fang, Ke Zhang, Mengling Yang, Aiting Jiang, and Xiongchao Yang, for their great contribution for the implementation and experimental studies. Without these tremendous supports, this book would not be finalized in due time.

The author apologizes in case of errors or mistakes in the book. The readers are welcome to provide feedback information to further improve the quality of the book.

This work was supported in part by the National Natural Science Foundation of China under Grant 51977026, in part by the Natural Science Foundation of Guangdong Province under Grant 2018A030313494, and in part by Sichuan Province Key Research and Development Project under Grant 2017GZ0347 and 2021YFG0255.

Chengdu, China   Yang Han

Contents

About the Author

Yang Han (S'08-M'10-SM'17) received his Ph.D. degree in electrical engineering from Shanghai Jiaotong University (SJTU), Shanghai, China, in 2010. He joined the University of Electronic Science and Technology of China (UESTC) in 2010, and has been an associate professor since 2013. From March 2014 to March 2015, he was a visiting scholar in the Department of Energy Technology, Aalborg University, Aalborg, Denmark. His research interests include the AC/DC microgrids, grid-connected converters for renewable energy systems and distributed generators (DGs), power quality, active power filters, and static synchronous compensators (STATCOMs).

Dr. Han was the recipient of the Provincial Science and Technology Award in 2020, Science and Technology Award from Sichuan Electric Power Company in 2019, Academic Talent Award by UESTC in 2017, and Baekhyun Award in2016 by the Korean Institute of Power Electronics. He was a session chair for Power Quality and Premium Power Supply Session at the 35th Annual Conference on Power System and Automation of Chinese Universities, Chengdu, China, in 2019, for Emerging Technologies and End-User Systems, Grid Operation and Management, and Power Electronics, Control and Protection Systems for Smart Grids Sessions in the IEEE PES Innovative Smart Grid Technologies Asia (ISGT Asia 2019), Chengdu, China, in 2019, for Microgrid and Distributed Generation Session in the Symposium on Power Electronics and

Electrical Drives (SPEED), Xi'an, China, in 2019, for
Microgrid Optimization and Scheduling Session in the
2nd International Conference on Power and Renewable
Energy, Chengdu, China, in 2017, for Power Quality
Mitigation and Application Session in the 5th National
Conference on Power Quality, Xi'an, in 2017, and for
AC/DC, DC/AC Power Converter Session in the 2016
IPEMC ECCE-Asia, Hefei, China.

Chapter 1
Introduction to the Modeling and Control of Power Electronic Converters for Microgrid Applications

1.1 Overview

With the increased penetration of distributed generation (DG) units on the electrical grid systems, the renewable energy sources (RESs) including micro-turbines, fuel cells, photovoltaic (PV) systems, and wind energy systems have been widely used in the distributed power systems in the past decades. The DG units play an important role in reducing pollution, decreasing power transmission losses, and improving local utilization of RESs, which becomes a strong support for the large-scale power grid. However, DG units may also bring challenges to the distribution network such as inverse power flow, voltage deviations, and voltage fluctuations. When a number of DG units are clustered together, they can form a microgrid (MG) that solves the problems caused by high penetration of DG units successfully and makes the large-scale application of DG systems possible [1–4].

Figure 1.1 shows the basic architecture of an AC MG system. The PV systems and energy storage systems (ESSs) are connected to the AC bus through the DC/DC/AC converters and wind turbines are tied to the AC bus through the AC/DC/AC converters. In the case of islanding operation, RESs mainly provide AC power to the loads through the local control. In the grid-connected mode, the AC MG is connected to the upstream grid through a tie line at the point of common coupling (PCC) and there is power flow between MG system and the grid [1, 2].

To understand the operating principle and design of a proper hierarchical controller, the mathematical modeling of voltage-source inverters (VSI) and controller design is crucially important. Hence, this book would focus on the modeling and control of single-inverter, closed-loop controller synthesis, stability analysis, and parameter selection criterion, and then extended to the case of multiple-inverter-based microgrid applications. Both linear and nonlinear modeling approaches for voltage-source inverters would be covered, and the well-known bode plot, Nyquist plot, pole-zero map would be used throughout this book, accompanied by the non-linear approach, such as the Jacobian matrix method and the Lyapunov exponent

© Springer Nature Switzerland AG 2022
Y. Han, *Modeling and Control of Power Electronic Converters for Microgrid Applications*, https://doi.org/10.1007/978-3-030-74513-4_1

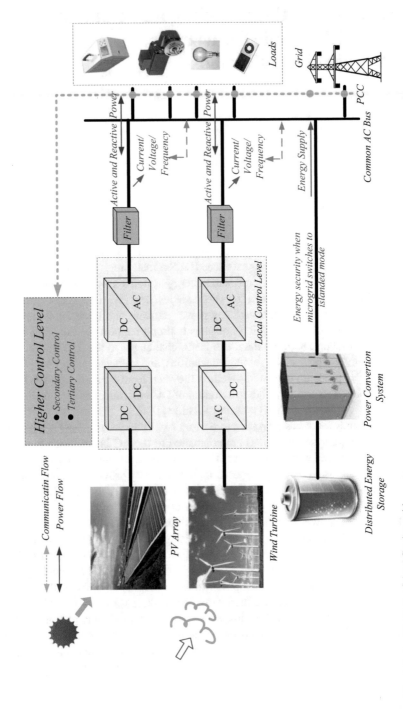

Fig. 1.1 Architecture of the AC microgrid

method, to explore the multi-time-scale characteristics of voltage-source inverters under closed-loop control. Furthermore, the hierarchical control architecture is applied for the microgrid system, where the droop control, secondary control, and the presented washout-filter-based control or consensus-based control approach are analyzed and compared.

As for the simulation case studies, the MATLAB/Simulink, PLECS, and Electromagnetic Transient Program - Alternative Transient Program (EMTP-ATP) are utilized throughout this book, and the extensive experimental studies are also given for validation purposes. Notably, the EMTP-ATP-based case studies are reported in most chapters, due to their open-source nature and the newly developed user-friendly interface ATPdraw. Another reason for the EMTP-ATP-based case studies is based on the teaching experience for the master course on electrical engineering simulation software in the past few years [5–8].

Figure 1.2 shows the newly developed ATPdraw user interface, and Fig. 1.3 shows the circuit diagram of a 50 MW PV power plant modeled in ATPdraw. Similar to the circuit model in Fig. 1.3, the voltage-source inverter including control algorithm and the microgrid system can also be easily modeled in the ATPdraw platform. For most of the applications, the power stage and controller model can be found in the newly developed ATPdraw library [5, 6]. However, in case of the missing components, the user-defined model using Fortran language, TACS, or MODEL language can be applied, especially for the advanced users. For detailed information regarding the Fortran, TACS, or MODEL language, one can refer to the official website of EMTP and ATPdraw [5, 6], thus it would be omitted here for the sake of brevity. Hence, this idea was applied for various case studies in the manuscript in the remainder of this book. Due to space limitations, detailed information about some case studies would be uploaded in the course website.

(Master Course Website: https://i.study.uestc.edu.cn/EES/menu/home)

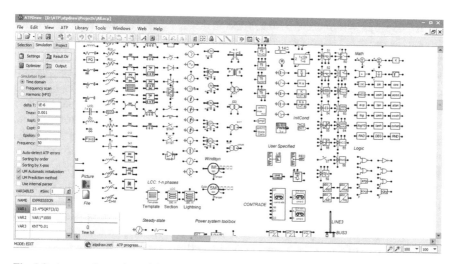

Fig. 1.2 A general overview of the ATPdraw software for time-domain simulation [5]

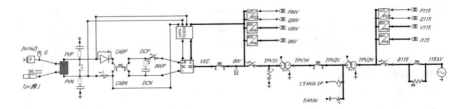

Fig. 1.3 The circuit diagram of a 50 MW PV power plant modeled in ATPdraw [5]

1.2 Scope of the Book

In Chap. 2, a tutorial on the parameter design of the *LCL*-filter is presented, as well as the modeling and stability analysis of the *LCL*-type grid-connected inverters. The generalized parameter design constraints of the *LCL* filter are introduced to facilitate passive component selection, and the magnetic integration techniques of filter inductors to reduce the weight and size of filter, and increase the power density of the inverter system. Moreover, the various damping methods for enhancing the individual internal stability and the relevant application issues are also outlined. Besides, the impedance-based method for evaluating system-level interactive stability is introduced, with an emphasis on the different modeling methods of inverter output impedance and online impedance measurement techniques. And then, the benchmark system for stability analysis of grid-connected LCL-type inverters is presented, where the impedance-based stability analysis approach is applied for stability evaluations using the proportional-resonant (PR) and synchronous reference frame proportional-integral (SRF-PI) controllers. Finally, future research trends on the modeling and stability analysis of *LCL*-type grid-connected inverters are also presented.

In Chap. 3, a systematic parameter design guideline for HRF-based $v + i_c$ control strategy is proposed to ensure system stability and optimize the performance of the system under control delay condition. The mathematic model of the HRF-based $v + i_c$ control strategy is established with the consideration of control delay. Based on this model, a satisfactory region of the system stability indexes can be obtained by stability specifications of the system, and the optimal control parameters can be calculated according to the stability indexes selected from the satisfactory region. The simulation results obtained from EMTP-ATP and the experimental results obtained from a reduced-scale prototype system are presented to validate the effectiveness of the optimal control parameter design methodologies, which can be widely applied for similar standalone inverters and uninterruptible power supply (UPS) systems.

In Chap. 4, the stability characteristics of a digital controlled single-phase voltage-source inverter (VSI) with SRF voltage control loop is investigated from the perspective of a nonlinear system. The stability analysis is implemented using the discrete-time model defined by the stroboscopic map, which is derived using the state-space averaging (SSA) technique. Furthermore, two different nonlinear analysis methods, the Jacobian matrix method and Lyapunov exponent method, are adopted to analyze the fast-scale stability and the slow-scale stability of the PWM inverter under variations of control parameters, hence the stability regions can be easily obtained analytically. Simulation results obtained from EMTP-ATP software are presented to study the effect of control delay, load parameters, and controller gain on the stability characteristics of the closed-loop system. In order to validate the theoretical analysis, the experimental results under resistive load, inductive-resistive load, and diode rectifier load conditions are presented, which also proves that the discrete-time model plus the Jacobian matrix method or Lyapunov exponent method is capable of accurately investigating the stability boundaries of a voltage-source converter operating in the standalone mode with SRF control loops.

In Chap. 5, the band-pass filter (BPF)-based droop control scheme is extended to an islanded single-phase microgrid in hybrid frame to achieve voltage amplitude and frequency deviation restoration. Moreover, the dynamic stability of the studied system is addressed by the derived reduced-order small-signal model, which simplifies the modeling process and theoretical analysis. Followed by the system model, the impact of system parameter variation on the stability and dynamic performance of the microgrid is subsequently predicted by applying the eigenvalue-based analysis approach. Finally, the effectiveness of eigenvalue analysis is verified by using extensive simulation results obtained from PLECS and EMTP-ATP, and the experiment results are provided to further validate the effectiveness of the BPF-based droop control method in the islanded single-phase microgrid.

In Chap. 6, the equivalence between a secondary control scheme and washout filter-based power sharing strategy for islanded microgrid is demonstrated, and the generalized washout filter control scheme is derived. And the physical meaning of control parameters of secondary controllers is also illustrated. Besides, a complete small-signal model of the generalized washout filter-based control method for an islanded MG system is built, which can be used to design the control parameters and analyze the stability of the MG system. Moreover, the simulation results obtained from EMTP-ATP are given to illustrate the difference between the conventional droop control scheme and the washout filter-based improved droop control scheme, and hardware-in-the-loop results are also presented to show a comparative analysis under generic operating conditions. Finally, the experimental results from a reduced-scale prototype system are provided to confirm the validity and effectiveness of the derived equivalent control scheme for a three-phase islanded MG.

In Chap. 7, the consensus-based enhanced droop control scheme is introduced for an islanded microgrid system, which achieves accurate active and reactive power sharing while maintaining the frequency recovery and keeping the average voltage to the rated values. In the proposed control scheme, only the neighborhood reactive power information needs to be exchanged by using a sparse low-bandwidth communication (LBC) network, instead of delivering information of active power, reactive power, and frequency by communication links in the existing consensus methods. Compared with the existing consensus-based methods, the transfer data and data latency are significantly reduced and high reliability of the system can be achieved. Moreover, the accurate active power sharing and frequency recovery can be ensured under disturbances of load and feeder impedance, even in case of communication failures. Finally, the steady-state performance and local exponential stability analysis of the proposed control scheme are also presented, and the simulation and hardware-in-the-loop (HiL) test results are provided for validation purposes.

In Chap. 8, the modeling, controller design, and stability analysis of the islanded microgrid (MG) using an enhanced hierarchical control structure with multiple current loop damping schemes is proposed. The islanded MG consists of parallel-connected voltage-source inverters using LCL output filters, and the proposed control structure includes primary control with an additional phase-shift loop, the secondary control for voltage amplitude and frequency restoration, the virtual impedance loops, and the inner voltage and current loop controllers. A small-signal model for the primary and secondary controls with an additional phase-shift loop is presented, and the moving average filter-based sequence decomposition method is proposed to extract the fundamental positive and negative sequences, and harmonic components. The multiple inner current loop damping scheme is presented, including the virtual positive, virtual negative, and variable harmonic sequence impedance loops for reactive and harmonic power sharing purposes and the proposed active damping scheme using capacitor current feedback loop of the LCL-filter, which shows enhanced damping characteristics and improved inner-loop stability features. The simulation results obtained from EMTP-ATP under non-identical line impedance scenario are presented, and the effect of the low-bandwidth communication (LBC) delay is also simulated and compared with the ideal scenario. Finally, the experimental results are also provided to validate the feasibility of the proposed approach, which can be widely applied in practical applications.

References

1. Han, Y., Li, H., Shen, P., Coelho, E. A. A., & Guerrero, J. M. (2017). Review of active and reactive power sharing strategies in hierarchical controlled microgrids. *IEEE Transactions on Power Electronics, 32*(3), 2427–2451.
2. Han, Y., Zhang, K., Li, H., Coelho, E. A. A., & Guerrero, J. M. (2018). MAS-based distributed coordinated control and optimization in microgrid and microgrid clusters: A comprehensive overview. *IEEE Transactions on Power Electronics, 33*(8), 6488–6508.

3. Han, Y., Shen, P., Zhao, X., & Guerrero, J. M. (2016). An enhanced power sharing scheme for voltage unbalance and harmonics compensation in an islanded AC microgrid. *IEEE Transactions on Energy Conversion, 31*(3), 1037–1050.
4. Han, Y., Lin, X., Fang, X., Yang, P., Hu, W., Coelho, E. A. A., & Blaabjerg, F. (2020). Floquet-theory-based small-signal stability analysis of single-phase asymmetric multilevel inverters with SRF voltage control. *IEEE Transactions on Power Electronics, 35*(3), 3221–3241.
5. www.atpdraw.net
6. www.emtp.org
7. www.plexim.com
8. www.mathworks.com

Chapter 2
Modeling and Stability Analysis of LCL-Filter-Based Voltage Source Inverters

With the increasing penetration of renewable energy to the utility system, the *LCL* filter has been widely adopted to interface between the inverter and the grid for improving the quality of injected grid currents due to the advantages of superior harmonics attenuation ability and reduced size. However, the high-order characteristics and various constraints of the *LCL* filter complicate the filter design process. Moreover, the stability of the internal current control loop of the individual inverter is susceptible to the inherent *LCL*-filter resonance. Meanwhile, the overall system stability would be deteriorated by the external interactions between the inverter and the weak grid, as well as among the paralleled inverters. Both the *LCL*-filter resonance and two types of interaction would cause severe distortion of the grid currents, which would deteriorate the power quality at the common coupling point (PCC) of the electric distribution system.

This chapter presents a tutorial on the parameter design of the *LCL*-filter, as well as the modeling and stability analysis of the *LCL*-type grid-connected inverters. The generalized parameter design constraints of the *LCL* filter are briefly introduced to facilitate the passive component selection, and the magnetic integration techniques of filter inductors to reduce the weight and size of the filter, and increase the power density of the inverter system. Moreover, the various damping methods for enhancing the individual internal stability and the relevant application issues are also outlined. Then, the impedance-based method for evaluating system-level interactive stability is introduced, with an emphasis on the different modeling methods of inverter output impedance and online impedance measurement techniques. Besides, the benchmark system modeled in EMTP-ATP for stability analysis of grid-connected LCL-type inverters is presented, where the impedance-based stability analysis approach is applied for stability analysis using the proportional-resonant (PR) and synchronous reference frame proportional-integral (SRF-PI) controllers. Finally, future research trends on the modeling and stability analysis of *LCL*-type grid-connected inverters are also presented.

© Springer Nature Switzerland AG 2022
Y. Han, *Modeling and Control of Power Electronic Converters for Microgrid Applications*, https://doi.org/10.1007/978-3-030-74513-4_2

2.1   Introduction

Figure 2.1 shows the diagram of the distributed power generation systems (DPGS), which have been widely utilized for renewable energy integration, such as solar, wind, and fuel cell, which greatly alleviate the energy crisis and environmental problems. Grid-connected inverters controlled by pulse-width modulation (PWM) techniques play a key role in promoting renewable energy consumption. However, the harmonics caused by the PWM process would impose additional challenges on the electric network, such as multiconverter resonance, converter-grid oscillation, or system destabilization induced by distorted grid current. The passive filters are usually connected between the grid and the inverters to attenuate the high-frequency harmonics to improve the quality of injected grid currents [1–5].

Normally, L filters are not suggested due to their poor high-frequency attenuation ability of -20 dB/dec and bulky inductors, yet LC filters are advantageous over L filters, with a harmonics attenuation rate of -40 dB/dec. Nevertheless, compared with L and LC filters, smaller inductors and capacitors are required in LCL filters, and the superior high-frequency attenuation characteristic of -60 dB/dec can also be achieved simultaneously. In this scenario, LCL-type grid-connected inverters are preferred to be adopted in practical applications.

To avoid the undesired stability problems resulting from the harmonics pollution, the filtering performance of LCL-filter can be maximized by means of an optimal filter design. Nonetheless, the high-order LCL filters complicate the parameter selection due to the contradiction among the parameters and various design constraints. Recently, lots of papers have been published to discuss the LCL-filter parameter design. Although the focuses are dissimilar in diverse design processes, some common principles still exist in different research works irrespective of the various applications, for instance, the selection of LCL-filter resonance frequency, the current ripple, the total inductance, the harmonic attenuation rate, and the reactive power absorbed by filter capacitor. It is worth mentioning that, although the inductances of the LCL filter are reduced compared with the inductances of L and LC filters, two discrete inductors are still redundant in view of the weight and volume of the LCL filter. In this scenario, the consideration of magnetic integration techniques in the filter design process is necessary to further minimize the bulky inductors. Note that the utilization of the LCL-type grid-connected inverters would

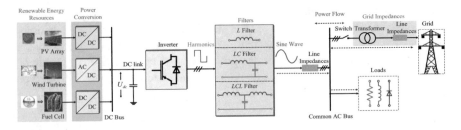

Fig. 2.1 DPGS with the LCL-filter interfaced grid-connected inverter [1–5]

result in additional stability issues even if the *LCL* filters are meticulously designed. Specifically, the stability of the internal current control loop of individual inverters itself is related to the inherent *LCL*-filter resonance peak, the so-called individual internal stability. Also, the overall system stability may deteriorate due to the underlying external interaction resonances between the inverter and the weak grid, as well as among paralleled inverters, namely, the external stability of the inverter [6–10].

In order to improve the internal stability of the individual inverters, the passive damping (PD) methods can be applied by adding resistors in series or parallel with the *LCL*-filter branches. However, the inevitable damping losses and degraded harmonics attenuation ability at high frequencies are yielded due to the presence of dissipated components. Furthermore, the complex PD methods are alternative to diminish the power losses and regain the filtering performance, yet the size and weight of filters are increased, arising from the additional passive components. On the other hand, by inserting a digital filter in the forward path of the current control loop, the filter-based damping method is also applicable to suppress the *LCL*-filter resonance peak, without extra sensors, whereas the system robustness is poor. In this case, an additional state variable can be fed back to damp the *LCL*-filter resonance and increase the system robustness, which emulates a physical resistor in the *LCL*-filter branch, that is, an virtual impedance. The analytical expression and connection type of virtual impedance are dependent on the feedback variable and coefficient, which has been extensively studied in recent literatures [11–18].

On the other hand, the external stability analysis approaches can be roughly categorized into state-space method in the time domain and impedance-based method in the frequency domain. Conversely, the impedance-based method can be used to evaluate the stability by exploring the terminal characteristics of the system, namely, whether the ratio of the inverter output impedance to the grid impedance satisfies the Nyquist stability criterion (NSC). Normally, the inverter output impedance can be derived by employing the equivalent transfer functions or small-signal linearization method, and can also be obtained by the impedance measurement technique by injecting the perturbation signals into the grid voltage and capturing the corresponding responses, in which case the system is regarded as a black box. Similarly, the grid impedance can be estimated by superimposing the small perturbations on the current reference signals, which can be identified in real time to predict the global stability of the interconnected system, to diminish the effect of time-varying characteristics of grid impedance.

This chapter presents a tutorial on the state-of-the-art techniques of *LCL*-type grid-connected inverters, including the *LCL*-filter design, and the internal and external stability of inverters. The remainder of this chapter is organized as follows. Section 2.2 reviews the generalized parameter design constraints and magnetic integration techniques of *LCL* filters. Subsequently, in order to solve the internal instability induced by the *LCL*-filter resonance peak, the damping methods, including the passive damping methods, filter-based damping methods and state-feedback-based damping methods are introduced in Sect. 2.3, and the influence of control delay on the system stability and the corresponding countermeasures are also discussed. Moreover, Sect. 2.4 gives an overview on the impedance-based stability analysis, impedance modeling methods, online impedance identification techniques, and the interactive stability analysis of the paralleled inverters. Section 2.5 presents the benchmark system for the stability analysis of grid-connected LCL-type inverters, where the impedance-based stability analysis approach is applied to evaluate the system stability using proportional-resonant (PR) and synchronous reference frame proportional-integral (SRF-PI) controllers. Finally, Sect. 2.6 concludes this chapter.

2.2   System Description

Figure 2.2 shows the equivalent circuit of an *LCL*-type grid-connected inverter system, where L_1 and L_2 are the inverter-side and grid-side inductors, respectively, C is the filter capacitor, Z_g is the grid impedance, i_1 and i_2 are the inverter-side and grid-side currents, respectively, i_C is the capacitor current, u_{inv} is the inverter output voltage, u_{pcc} is the voltage at PCC, u_C is the capacitor voltage, and u_g is the grid voltage. As an interface between the inverter and the grid, the *LCL* filter improves the quality of injected grid current and voltage at the point of common coupling (PCC), thus avoiding converter-grid oscillation or even destabilization caused by harmonic pollution issues. Specifically, the preferred properties of *LCL* filter include high-current ripple rejection capabilities, fast dynamic response, low voltage drop, high power factor, and low volume and weight. Meanwhile, magnetic integration techniques are crucial for reducing the device volume or constructing the higher-order output filters.

2.2.1   Parameter Design Procedures

This subsection summarizes the procedure for parameter selection, including selection of filter capacitor C, the total inductance L_T, the inverter-side inductance L_1, the harmonic attenuation rate δ, and the resonance frequency f_r [6, 8, 16].

- *Filter Capacitor Selection:*
Since the capacitor branch is the dominant flow path of high-frequency current harmonics. The selection of capacitor value should achieve a tradeoff between the

Fig. 2.2 The equivalent circuit of the *LCL*-type grid-connected inverter system

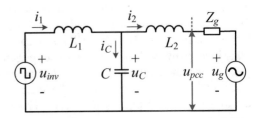

power factor (PF) and the harmonics attention ability of the *LCL* filter and, therefore, the reactive power stored in the capacitor is usually less than 5% of the rated active power of the inverter.

- *Total Inductance Selection*

With respect to the total inductance L_T, the fundamental voltage drop across the filter inductors should be less than 10% of grid voltage. Note that the high dc-link voltage is needed to assure the current controllability in the case of large L_T, which leads to high switching losses.

- *Inverter Side Inductance Selection*

Normally, the allowable maximum current ripple is appropriately selected for obviating the inductor saturation problems, and the L_1 is inversely proportional to maximum current ripple. In addition, the resonance frequency f_r should be in the range of $10f_0 \leq f_r \leq f_{sw}/2$, where f_0 is the fundamental frequency of grid voltage and f_{sw} is the switching frequency.

Table 2.1 summarizes the generalized design constraints to facilitate the parameter selection of the *LCL* filter, where U_{rms} is the RMS of fundamental line-to-line grid voltage, P_n is the rated active power of the inverter, U_{dc} is the dc-link voltage, I_{rated} is the rated current of the inverter, i_{1sw} and i_{2sw} are the switching frequency current harmonics across L_1 and L_2, respectively, and ω_{sw} is the switching angular frequency.

In general, the initial conditions are determined ahead of the design process, including $U_{rms}, P_n, f_{sw}, f_0, I_{rated}$. Then the parameter selection of the *LCL* filter can be realized by using the following step-by-step design procedure [1, 5, 6].

Step 1: Determine the maximum total inductance L_{Tmax}.

Table 2.1 A summary of constraints for choosing *LCL*-filter parameters [6, 8, 16]

Parameters of the LCL filter	Constraints		Impact of parameters on filter performance
Filter capacitor C		$C < 5\% \dfrac{P_n}{2\pi f_0 U_{rms}^2}$	• Large C results in low power factor • Small C requires large inductance
Total inductance L_T ($L_T = L_1 + L_2$)		$L_T \leq 10\% \dfrac{U_{rms}^2}{2\pi f_0 P_n}$	• Large L_T results in large voltage drop, small current ripple and PF, poor dynamic response and high cost
Inverter-side inductance L_1	1-phase inverter with unipolar SPWM, $r = 8$	$L_1 \geq \dfrac{U_{dc}}{(20\% \sim 30\%)I_{rated} r f_{sw}}$	• Large L_1 results in small current ripple, high voltage drop and cost
	1-phase inverter with bipolar SPWM, $r = 2$		
	Three-phase two-level inverter using SPWM or SVM, m_i is the modulation index	$L_1 \geq \dfrac{\sqrt{3}}{12} \dfrac{U_{dc}}{30\% I_{rated} f_{sw}} m_i$	
Harmonic attenuation rate δ $\delta = \left\| \dfrac{i_{2sw}}{i_{1sw}} \right\| = \dfrac{1}{\|1 + n_L(1 - L_1 C \omega_{sw}^2)\|}$, $\omega_{sw} = 2\pi f_{sw}$		$\delta = 20\%$	• Small δ corresponds to low THD
Resonance frequency f_r $\omega_r = 2\pi f_r = \sqrt{(L_1 + L_2)/L_1 L_2 C}$		$10f_0 < f_r < 0.5 f_{sw}$	• Small f_r results in narrow control bandwidth • Large f_r results in resonance peak near f_{sw}

Step 2: Select L_1 greater than the minimum inverter-side inductance L_{1min}.

Step 3: Determine the maximum capacitor value C_{max}. The initial C can be chosen as one half of C_{max} to reduce the iteration design process.

Step 4: Select the original harmonic attenuation rate δ as 20% for a tradeoff between the total harmonic distortion (THD) of grid current and filter costs. Then the L_2 can be determined according to the δ and the ratio a_L between L_2 and L_1, that is, $a_L = L_2/L_1$. Besides, the L_2 can also be directly chosen by setting $a_L = 1$, in which case the size of the passive components can be minimized, whereas $a_L = 1$ corresponds to the minimum harmonic attention. Note that if the sum of L_2 and L_1 is greater than the L_{Tmax}, a new design procedure should be easily implemented by increasing C.

Step 5: Verify that the resonance frequency f_r is within the reasonable range. If f_r is lower than $10f_0$, the capacitor value in Step 3 should be reduced. Conversely, if f_r is higher than $f_{sw}/2$, the capacitor value or f_{sw} should be increased.

Step 6: Check that the THD of grid current is lower than 5%. If the THD is higher than 5%, the δ should be properly reduced with a new design procedure [5, 6].

Therefore, the parameter selection of the *LCL* filter is an iteration process until all the constraints are satisfied. Next, the final parameters of the *LCL* filter can be determined, then the magnetic design of the inductors needs to be considered in the next design procedure, such as core materials and size, winding turns, air gaps and core shapes.

2.2.2 Magnetic Integration Techniques

To reduce the weight and size of filters, two discrete inductors can be integrated into one magnetic core, where the magnetic flux of the two inductor windings in the common path is reversed to reduce the total magnetic flux of common core and, therefore, the cross-section area of common core is diminished, thereby decreasing the volume of the overall magnetic core. Conversely, the magnetic flux direction of two windings in the common core can also be identical, in which case the coupling effect between L_1 and L_2 can be intentionally maximized to construct the equivalent high-order filters, without any extra trap inductors. The magnetic integration techniques are mainly related to the selection of the magnetic materials, core shapes, adopted wires, and winding methods [17–20].

Notably, soft magnetic materials, such as ferrites, laminated silicon steel, powder core, amorphous alloys, and nanocrystalline materials, are generally utilized in the filter inductors of power electronic devices. It is noteworthy that high saturation flux density and relative permeability of magnetic materials contribute toward diminishing the winding turns of inductors, and high Curie temperature guarantees the stable inductance. Specifically, the ferrite materials exhibit low core losses and saturation flux densities, which are unsuitable for the large-current high-power applications.

Moreover, thin laminated silicon steel is commonly used on account of the low eddy current losses and lighter weight. The powder core is the superior material owing to its low eddy current losses and stable inductance against the temperature variation. However, the distributed air gap in the powder core may cause additional fringing effect losses, and their nonlinear current-dependent inductance characteristics would affect the effective *LCL*-filter design. Also, the amorphous alloys and nano-crystalline are alternative materials due to the low core losses and high saturation magnetic flux densities, yet the high cost may be unacceptable in practical applications. Actually, the utilization of copper foils and Litz wires can further reduce the copper losses caused by skin effects.

Figures 2.3, 2.4, and 2.5 show the different core structures and the corresponding equivalent circuits, including the EIE-, UIU-, and EE-type cores, where Φ_1, Φ_2, and Φ_c are the magnetic fluxes generated by inverter-side current i_1, grid-side current i_2, and capacitor current i_C, respectively, the air-gaps (g_1 and g_2) and winding turns (N_1, N_2, and N_c) can be adjusted to obtain desired inductances. Figure 2.3 shows the integrated EIE-type core structure, which can be used to reduce the volume and weight of filter, where the positive coupling effect between L_1 and L_2 introduces a negative mutual inductance $-M_{12}$ in the capacitor branch. Consequently, the high-frequency attenuation slope of the integrated *LCL* filter is simultaneously degraded due to the $-M_{12}$. In this way, the properly decreased coupling coefficient can be employed to ensure effective filtering performance. In order to counteract the $-M_{12}$, an active magnetic decoupling winding on the I-type core can be applied in the UIU-type core structure, that is, $L_c + 2M_{1c} = M_{12}$, as shown in Fig. 2.4, where L_c is the self-inductance of the decoupling winding, and M_{1c} is the mutual inductance. Moreover, by integrating two inductors into the EE-type core structure with negative coupling effect between inductor windings, a positive mutual inductance M_{12} can be introduced in the capacitor branch to construct an equivalent *LLCL* filter, as shown in Fig. 2.5. Then, strong switching-frequency harmonics attenuation can be

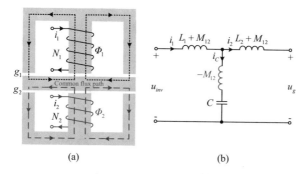

(a)   (b)

Fig. 2.3 Integrated *LCL* filter with positive coupling effect [18, 19]. (**a**) EIE-type core structure. (**b**) Equivalent circuit

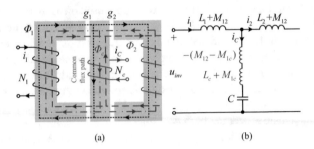

Fig. 2.4 Active magnetic decoupling for the positive coupled *LCL* filter [20]. (**a**) UIU-type core structure. (**b**) Equivalent circuit

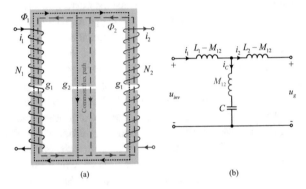

Fig. 2.5 Constructed *LLCL* filter with negative coupling effect [17]. (**a**) EE-type core structure. (**b**) Equivalent circuit

achieved. However, the double volume and weight of common flux path are required to avoid magnetic saturation.

Table 2.2 summarizes the specific design aspects about inductor selection of the LCL filters. To conclude, the positive coupling effect between the inductor windings should be significantly reduced for improving the filtering performance of the *LCL* filter, which can be realized by decreasing the coupling coefficient of windings. Conversely, in order to construct an equivalent high-order filter, the negative coupling effect between L_1 and L_2 needs to be effectively maintained for introducing a positive mutual inductance in the capacitor branch. Indeed, from the above discussion on the parameter selection and inductors integration of the *LCL* filter, the quality of injected grid current can be improved while increasing the power density of the system, with reasonable filter design. Under this scenario, the implicit menace caused by PWM to the conventional grid can be eliminated, namely, obviating the grid oscillation or even destabilization induced by the harmonic pollution. However, the instability of an individual inverter induced by the inherent LCL-filter resonance peak cannot be exempted irrespective of the deliberate filter design and, therefore, additional countermeasures are naturally required to damp the potential resonance effect.

Table 2.2 Specific aspects about inductors selection of *LCL* filters [7, 17–20]

Specific aspects		Advantages	Disadvantages	Applications
Core materials	Ferrites	• Low cost • Low core losses • High resistivity	• Fragile • Low saturation flux density • Low Curie temperature	• Low-current applications
	Laminated silicon steel	• Low eddy current losses • Reduced acoustic noise • High Curie temperature	• Fragile • Reduced saturation flux density • Reduced life time of laminated materials	• Low-current applications
	Powder core	• Low eddy current losses • High saturation flux density • High resistivity and Curie temperature	• Cause fringing effect losses • Electromagnetic interference	• High-current applications
	Amorphous alloys	• Low core losses • High saturation flux density	• Low Curie temperature • High cost	• High-current applications
	Nanocrystalline materials	• Low eddy current losses • High saturation flux density • High Curie temperature	• High cost • Cause fringing effect losses	• High-current applications
Core shapes and Winding methods	EIE-type with positive coupling	• Reduced volume and weight • Reduced core losses	• Degraded high-frequency attenuation ability	• Integrate single-phase or three-phase *LCL* filter
	EE-type with negative coupling	• Strong attenuation ability for switching-frequency harmonics	• Poor attenuation ability away from the switching frequency • Lack of design flexibility	• Construct equivalent *LLCL* filter
	UIU-type with positive coupling	• Ensure harmonics attenuation ability of -60dB/dec	• Increased windings • Stringent design requirement	• Decoupling the integrated *LCL* filter
Adopted wires	Copper foils	• High space utilization • Small leakage inductance	• High cost • Difficult manufacturing process	• High-current applications • Less winding turns
	Litz wires	• Easy winding • Low eddy current losses	• Poor overload capability	• Low-current applications • More winding turns

Fig. 2.6 The placements of damping resistor of six typical PD methods [8, 21]

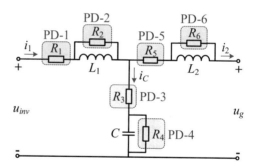

2.3   Damping Methods for Internal Stability

Since the stability of the internal current control loop of an individual inverter would be aggravated due to the *LCL*-filter resonance, the various damping methods can be employed to increase system damping for solving the resonance problem, including the passive damping (PD) methods, the filter-based damping methods, and the state-feedback-based damping methods.

2.3.1   Passive Damping Methods

Figure 2.6 shows the six typical PD methods to suppress the *LCL*-filter resonance, which can be utilized by adding the series or paralleled resistors into the *LCL* filter branches. The R_1, R_3, and R_5 are the damping resistors in series with L_1, C, and L_2, respectively, and R_2, R_4, and R_6 are the damping resistors in parallel with L_1, C, and L_2, respectively. Moreover, R_1 and R_5 can be regarded as the equivalent resistances of L_1 and L_2, respectively. Figure 2.7 shows the Bode diagrams of six typical PD methods, and the transfer functions corresponding to the various damping methods are also incorporated.

Corresponding to the PD methods in Fig. 2.7b, c, and f, the high-frequency attenuation slope is reduced in comparison with that of the undamped *LCL* filter since the additional zeros are contained in the transfer functions. In Fig. 2.7a, e, the attenuation ability of high-frequency harmonics is unaffected, whereas the large damping losses are caused by PD-1 and PD-5 owing to the directed path of the power flux through R_1 and R_5. Meanwhile, the utilization ratio of DC voltage and the dynamic tracking performance are weakened due to the presence of diminished low-frequency gain and, therefore, PD-1 and PD-5 are not recommended. Besides, the comparatively large resistances are required in PD-2 and PD-6 to achieve PD, with decreased harmonics attenuation ability. Notably, the filtering performance of PD-4 is the best among the six PD methods, with invariable frequency characteristics, yet the damping losses are relatively high due to the effect of PCC voltages.

From the perspective of power losses, effective damping, and filtering performance, the PD-3 using a small resistance is preferred in comparison with other PD methods, despite the degraded high-frequency harmonics attenuation ability.

Table 2.3 summarizes a comparison of the typical PD methods and the selection conditions of the damping resistances, where ω_r is the resonance angular frequency of the *LCL*-filter, k_p is the proportional coefficient of the current controller, and K_{PWM} is the gain of the voltage source inverter. Obviously, the power losses caused by PD-3 are inevitable no matter in high or low power applications. Based on PD-3, several complex PD methods can be used to reduce a certain amount of damping losses while retaining high-frequency harmonics attenuation ability, which are realized by adding the shunt capacitor or inductor in the capacitor branch, as indicated by the passive elements marked in red in Fig. 2.8.

In Fig. 2.8a, the power losses caused by the fundamental current are minimized owing to the low impedance path, that is, $\omega_0 L_d \ll R_d$. In addition, the damping resistor R_d should be the dominant path of resonance current to increase the system damping, that is, $R_d \ll \omega_r L_d$. Based on Fig. 2.8a, the additional C_d in Fig. 2.8b is employed to decrease the power losses caused by the switching-frequency current harmonics, and the conditions of $R_d \gg 1/(C_d \omega_{sw})$ and $R_d \ll 1/(C_d \omega_r)$ should be satisfied for low damping losses and suitable damping, respectively. Besides, the high-frequency harmonics attenuation slope of the filter shown in Fig. 2.8c is still -60 dB/ dec, yet two possible resonance peaks may be induced with the variation of R_d. Moreover, the total capacitance in Fig. 2.8c should be consistent with the capacitance of the undamped *LCL* filter, and a tradeoff between the damping performance and the power losses can be attained when the condition $C/C_d = 1$ is satisfied. In Fig. 2.8d, the fundamental and switching frequency components of currents are bypassed by L_d and C_d, respectively and, therefore, the damping losses on R_d are minimum attributed to the only minor resonance current through R_d. Furthermore, by tuning the $L_d - C_d$ branch at the switching frequency in Fig. 2.8e, the switching

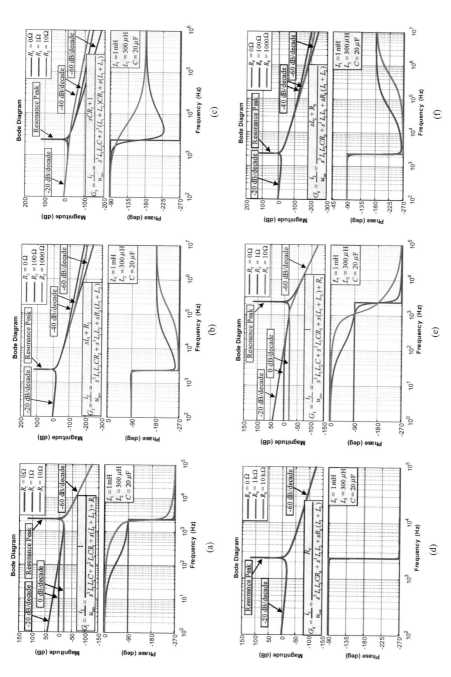

Fig. 2.7 Bode diagrams of six typical passive damping methods [8]. (**a**) PD-1. (**b**) PD-2. (**c**) PD-3. (**d**) PD-4. (**e**) PD-5. (**f**) PD-6

Table 2.3 Comparison of the typical passive damping methods for the LCL filters [21, 22]

Damping methods	Damping resistances selection	Compared with the undamped *LCL* filter	
		Merits	Drawbacks
PD-2	$0 < R_2 < \dfrac{(L_1+L_2)+\sqrt{(L_1+L_2)^2+4K_{PWM}^2k_p^2L_1C}}{2K_{PWM}k_pC}$	• Low-frequency gain is unaffected	• Degraded high-frequency harmonics attenuation ability • Cause damping losses
PD-3	$\dfrac{1}{6\pi L_1}\dfrac{L_2 f_{sw}}{f_r}\dfrac{1}{C\omega_r} \leq R_3 \leq \dfrac{1}{2\pi f_{sw}C}$	• Low-frequency gain is unaffected • Only need a small damping resistor	• Degraded high-frequency harmonics attenuation ability • Cause damping losses
PD-4	$0 < R_4 < \dfrac{L_1+L_2}{K_{PWM}k_pC}$	• Frequency characteristics are unaffected • Superior damping performance	• Large damping losses • Poor ability of reference tracking and disturbance rejection
PD-6	$0 < R_6 < \dfrac{(L_1+L_2)+\sqrt{(L_1+L_2)^2+4K_{PWM}^2k_p^2L_2C}}{2K_{PWM}k_pC}$	• Low-frequency gain is unaffected • Fast dynamic response • Superior disturbance rejection ability	• Degraded high-frequency harmonics attenuation ability • Large damping losses

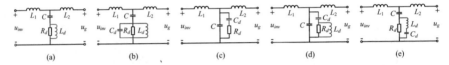

Fig. 2.8 The complex PD methods [15, 22]. (**a**) $C + (R_d//L_d)$. (**b**) $C + (C_d//R_d//L_d)$. (**c**) $C // (C_d+R_d)$. (**d**) $C // (C_d+R_d//L_d)$. (**e**) $C + (C_d+L_d)// R_d$

current harmonic is mostly bypassed due to the introduced low impedance path and, thus, the power losses are significantly diminished.

Table 2.4 shows a comprehensive comparison among the complex PD methods and the PD-3, and the parameter selection conditions of passive components are also summarized.

To conclude, the fundamental and the switching-frequency currents through the damping resistor can be bypassed to reduce the power losses. In comparison, the power losses of the complex PD methods in Fig. 2.8b, d are the lowest due to the maximum bypass of current components. Moreover, the approach with only an additional capacitor in Fig. 2.8c is easiest to be implemented among these complex PD methods, since the consideration of the complicated inductor design is not required in this case. However, the overall complexity of the circuit topologies and parameters design, and the cost and volume of *LCL*-filter are increased with the adoption of complex PD methods, yet the damping losses cannot be eliminated completely. In this case, the methods for increasing system damping by modifying the control algorithms have become crucially important, without any passive components and power losses.

Table 2.4 Performance comparison and parameters selection of the different complex passive damping methods [15, 22, 23]

Performance compared with PD-3 and parameters selection		$C + (R_d//L_d)$	$C + (C_d//R_d//L_d)$	$C // (C_d+R_d)$	$C // (C_d+R_d//L_d)$	$C + (C_d+L_d)// R_d$
Parameters selection		$\dfrac{R_d}{L_d\omega_0}=\dfrac{L_d\omega_r}{R_d}$	$1/(C_d\omega_r)=\dfrac{R_d}{1/(C_d\omega_{sw})}$	$r_c\sqrt{\dfrac{1}{rc}+1}\sqrt{\dfrac{L_2}{C}}\le R_d \le(1+r_c)\sqrt{\dfrac{L_2}{C}}$, $L_p=L_1L_2/(L_1+L_2)$, $r_c=\dfrac{C}{C_d}=1$ $R_d=\sqrt{(L_1+L_2)/((C+C_d))}$, $C=C_d$	$R_d=\sqrt{(L_1+L_2)/(C+C_d)}$, $L_d=2R_d\sqrt{(L_1//L_2)C}$, $C=C_d$	$R_d=\dfrac{1}{2\pi f_r C}$, $f_{sw}=\dfrac{1}{2\pi\sqrt{L_dC_d}}$
Slope of high-frequency attenuation		-40 dB/dec	-60 dB/dec	-60 dB/dec	-60 dB/dec	-40 dB/dec
Power losses caused by various components	Fundamental current	↓	↓	—	↓	—
	Resonance current	—	—	—	—	—
	Switching harmonics	—	↓	—	↓	↓
Harmonics attenuation of various components	Resonance current	—	—	—	—	—
	Switching harmonics	—	↑	↑	↑	↑

Note: The symbols ↑, ↓ and — represent the increase, decrease and unchanging, respectively

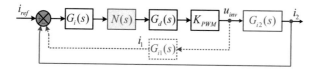

Fig. 2.9 Control diagrams of the filter-based damping methods with different current feedback strategies (blue dotted line represents the ICF, and red solid line represents the GCF) [9]

2.3.2 Filter-Based Damping Methods

By inserting a digital filter with special function in the forward path of the current control loop, the *LCL*-filter resonance peak can be damped by using the filter-based damping method, without any additional sensors and passive components. Normally, the filters mainly include the notch filter, the low-pass filter (LPF), and the all-pass filter.

Figure 2.9 shows the control block diagram of the filter-based damping methods, and the close-loop control scheme for reducing the steady-state error can be either inverter-side current feedback (ICF) or grid-side current feedback (GCF). The digital filter represented by $N(s)$ is usually cascaded to the current controller $G_i(s)$ in the forward path. The $G_i(s)$ is generally a proportional integral (PI) or a proportional resonant (PR) controller, which is employed in *dq*- or stationary frame to track the current reference i_{ref} without steady-state error, respectively; $G_d(s)$ is the digital control delay, and K_{PWM} is the gain of inverter bridge. The $G_{i1}(s)$ and $G_{i2}(s)$ are the transfer functions from inverter output voltage u_{inv} to inverter-side current i_1 and grid-side current i_2, respectively, which are given as

$$G_{i1} = \frac{i_1}{u_{inv}} = \frac{1}{sL_1} \frac{s^2 + \omega_a^2}{s^2 + \omega_r^2} \tag{2.1}$$

$$G_{i2} = \frac{i_2}{u_{inv}} = \frac{1}{sL_1} \frac{\omega_a^2}{s^2 + \omega_r^2} \tag{2.2}$$

where ω_a is the antiresonance angular frequency introduced by ICF, and ω_r is the resonance angular frequency of *LCL* filter, expressed as

$$\omega_a = 2\pi f_a = \sqrt{\frac{1}{L_2 C}} \tag{2.3}$$

$$\omega_r = 2\pi f_r = \sqrt{\frac{L_1 + L_2}{L_1 L_2 C}} \tag{2.4}$$

1. Notch Filter

Specifically, the typical transfer function of the notch filter is denoted as follows:

$$N(s) = \frac{s^2 + \omega_n^2}{s^2 + qs + \omega_n^2} \tag{2.5}$$

where q is the quality factor of notch filter, and ω_n is the notch angular frequency, $\omega_n = 2\pi f_n$.

Figure 2.10 shows the Bode diagrams of ICF and GCF schemes, with and without the notch-filter-based damping method, respectively. In Fig. 2.10a, without

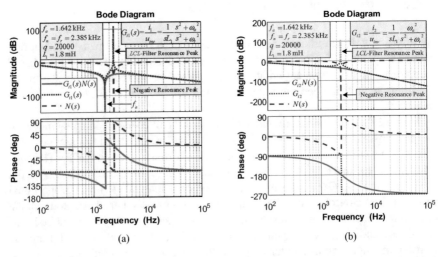

Fig. 2.10 Bode diagrams of notch-filter-based damping method with different current feedback strategies [24]. (**a**) ICF. (**b**) GCF

considering the control delay, the system using ICF is inherently stable, whereas the system using GCF scheme in Fig. 2.10b is unstable due to the only downward $-180°$ crossing with the gain above 0 dB. After employing the notch filter-based damping method, the positive LCL-filter resonance peak is counteracted by the notched resonance peak, with the condition $f_n = f_r$. Actually, this method is an intrinsical zero-pole cancellation of the system transfer function, that is, the zero of notch filter and the unstable pole of the LCL filter, which attenuates the resonance peak of the magnitude-frequency curve of the system. Note that a small q corresponds to a narrow rejection bandwidth, which results in high sensitivity to the deviation of the resonance frequency.

2. Low-Pass Filter

The selection of the cutoff frequency of the low-pass filter (LPF) is a tradeoff between the control bandwidth and the stability margin of the system. Generally, the cutoff frequency can be chosen near to the LCL-filter resonance frequency. It is noteworthy that the phase lag caused by LPF contributes to provide proper damping for the system with GCF, and the essence is to shift the phase-frequency curve of the system outside of the unstable frequency region, thereby stabilizing the whole system. The transfer function of the second-order LPF is expressed as

$$N(s) = \frac{\omega_r^2}{s^2 + 2D\omega_r s + \omega_r^2} \tag{2.6}$$

where ω_r is the resonance angular frequency of LCL filter, and D is the damping coefficient, usually selected as $1/\sqrt{2}$.

3. All-Pass Filter

Similarly, the proper phase lag introduced by an all-pass filter can also be utilized to enlarge the phase margin of the system at f_r. However, the dynamic performance and robustness of the system are degraded in this scenario. The all-pass filter can be described by Eq. (2.7).

$$N(s) = e^{j\phi_d} \tag{2.7}$$

where ϕ_d is the phase lag introduced by all-pass filter.

Table 2.5 summarizes the merits and drawbacks of different filter-based damping methods, which can be categorized into the approaches based on zero-pole cancellation and phase lag in view of the implementation principles. According to the NSC, the former is to avoid the gain above 0 dB at the frequency of $-180°$ crossing, and the latter is to obviate the $-180°$ crossing in the frequency range with a gain greater than 0 dB. It is worth noting that, the damping methods based on the phase lag filters are normally suitable for the GCF, whereas the dynamic performance of the system is degraded in this case due to the reduced control bandwidth. Conversely, the damping of the system with ICF can be increased by using the phase lead filters, in which case the problem of phase overcompensation needs to be precluded. In

Table 2.5 Merits and drawbacks of different filter-based damping methods [9, 23, 24]

Adopted filters	Merits	Drawbacks	Major technologies
Notch filter	• Simple implementation • Superior damping performance	• Sensitive to the variation of f_r • Small phase margin at low frequencies	• Select the f_n as f_r • Know the resistances of inductors for tuning process
Low-pass filter	• Stabilize system with i_2 feedback at low resonance frequency • Hardly dependent on the grid parameters	• Cutoff frequency limits the control bandwidth • If f_c is close to the closed-loop bandwidth, resonance damping is deficient	• Select the cutoff frequency as f_r • Proper phase lag
All-pass filter	• High-frequency noise is not amplified • Simplify current controller design	• Degraded transient performance (first-order all-pass filter) • Reduced system robustness (second-order all-pass filter)	• Maintain zero open-loop phase at f_r

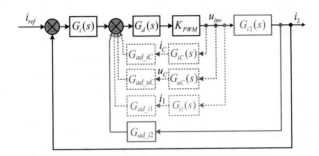

Fig. 2.11 Control block diagram of various state-feedback-based damping methods [25]

comparison, the notch filter is more effective due to the strong suppression perfor-mance with respect to the resonance peak. The methods based on the phase lead or phase lag are relatively complicated to implement, due to the stringent design requirement about that the degree of phase lead or lag. Although the filter-based damping methods show benefits of low cost and simple implementation, their robustness is subject to the variation of *LCL*-filter resonance frequency. Therefore, an additional state feedback is generally adopted to improve system robustness.

2.3.3   State-Feedback-Based Damping Methods

In addition to the original closed-loop control variables used to reduce the steady-state error of the system, an extra state variable can also be fed back to increase system damping. The supplemental feedback variable for increasing the system damping can be the capacitor current feedback (CCF), the capacitor voltage feed-back (CVF), the ICF, or the GCF. The control block diagram of the system with different states feedback is illustrated in Fig. 2.11.

In Fig. 2.11, the G_{ad_iC}, G_{ad_uC}, G_{ad_i1}, and G_{ad_i2} are the feedback coefficients to increase damping actively. The $G_{iC}(s)$ and $G_{uC}(s)$ are the transfer functions from inverter output voltage u_{inv} to i_C and u_C, respectively, which are expressed as

$$G_{iC} = \frac{i_C}{u_{inv}} = \frac{1}{sL_1} \frac{s^2}{s^2 + \omega_r^2} \tag{2.8}$$

$$G_{uC} = \frac{u_C}{u_{inv}} = \frac{1}{L_1 C} \frac{1}{s^2 + \omega_r^2} \tag{2.9}$$

Note that the state-feedback-based damping method is equivalent to a virtual impedance in series or parallel with the LCL-filter branch, the so-called virtual impedance-based method. Thereafter, the different states' feedback and their corresponding virtual impedances are discussed as follows.

1. Capacitor Current Feedback Method

In Fig. 2.11, the system can be stable with increased resonance damping when the G_{ad_iC} is a proportional coefficient. In this scenario, the CCF is equivalent to a virtual impedance in parallel with the filter capacitor. The virtual impedance in the case of proportional feedback can be modeled as follows:

$$Z_{VI}(s) = R_{VI} // (jX_{VI}) = \frac{L_1}{K_{PWM} C G_{ad\_iC} G_d(s)} \tag{2.10}$$

Specifically, the resonance frequency would be shifted due to the virtual reactance X_{VI}, and the LCL-filter resonance peak is damped, attributed to the virtual resistor R_{VI}. Yet, the damping method is invalid when the R_{VI} is negative, which can be solved by adopting the PI feedback or the first-order HPF feedback of i_C. Moreover, the resonance damping can also be achieved when the G_{ad_iC} is a second-order HPF with R_{VI} gain, in which case the CCF can be equivalent to a virtual pure resistor R_{VI} in series with the filter capacitor. However, the high-frequency noise would be easily amplified by the approximate derivative characteristics of HPF, and the control algorithm is more complicated than that of proportional feedback.

2. Capacitor Voltage Feedback Method

In addition, the resonance damping can also be yielded by applying the first-order derivative feedback of capacitor voltage since $i_C(s) = sCu_C(s)$, which emulates a physical impedance in parallel with the filter capacitor. However, the direct derivative operation is difficult to be implemented in digital controllers. Hence, the indirect derivation, such as the HPF and the lead-lag network can be adopted to approximate the direct derivation. Nonetheless, a large phase error introduced by the HPF leads to insufficient derivation at high frequency, and the lead-lag network is only suitable for a strong grid. The better indirect derivation schemes are the nonideal generalized integrator (GI) and the quadrature-second-order generalized integrator (Q-SOGI), which avoid the problem of noise amplification at high frequencies. Nevertheless, practical application is limited due to the complex algebraic process of GI and low robustness of Q-SOGI.

3. Inverter-Side Current Feedback Method

Generally, the ICF is utilized as the overcurrent protection for the inverter. As for increasing system damping, the proportional ICF is feasible to simulate an equivalent virtual impedance in series with L_1. Nonetheless, the additional resonance might be excited between C and L_2 due to the grid voltage harmonics, and the dynamic tracking performance of this strategy is also degraded because of the decreased low-frequency gain. Furthermore, the inverter-side current and the grid-side current can be collaboratively controlled by means of weighted average control (WAC), which simplifies the system order from third order to first order in this case. The control block diagram of the WAC is illustrated in Fig. 2.12, where β and $(1 - \beta)$ are the weight values of i_1 and i_2, respectively, and i_{WA} is the weighted average current. The coefficient β is defined by (2.11).

$$\beta = \frac{L_1}{L_1 + L_2 + L_g} \tag{2.11}$$

where L_g is the grid inductance. Apparently, the robustness of WAC is susceptible to the variation of grid inductance.

Without considering the effect of control delay, the WAC of the currents can be regarded as a virtual impedance in parallel with the filter capacitor, in which case the high-pass characteristics of the virtual impedance enable a low-impedance flow path for high-frequency harmonics equivalently, and the virtual impedance is modeled as follows:

$$Z_{VI}(s) = \frac{L_1}{C(1-\beta)G_i(s)} \tag{2.12}$$

In addition, the ICF scheme can also be regarded as a virtual impedance in series with L_1 when G_{ad_i1} is a first-order HPF. Hence, the expression of virtual impedance is given as

$$Z_{VI}(s) = R_{VI} + jX_{VI} = K_{PWM}G_{ad\_i1}(s) \tag{2.13}$$

4. Grid-Side Current Feedback Method

With respect to the GCF, the main merit of this strategy is that only one current sensor is required for both injected grid current tracking and resonance damping.

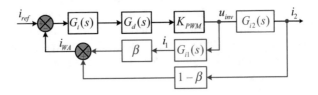

Fig. 2.12 Control block diagram of the weighted average control [26, 27]

Notably, the *LCL*-filter resonance peak can be suppressed by applying the second-order derivative feedback of i_2. In order to avoid the difficulty in the implementation of direct derivation, a first-order HPF with a negative gain can be adopted in the additional GCF damping loop. However, the control bandwidth of the system is narrowed to some extent, and the dynamic performance is poor for the system with low resonance frequency. Note that the first-order HPF-based damping method is equivalent to a virtual impedance in parallel with the filter capacitor, as follows:

$$Z_{VI}\left(s\right) = R_{VI} /\!/ \left(jX_{VI}\right) = \frac{L_1 L_2 s^2}{G_{ad\_i2}\left(s\right) K_{PWM}} \tag{2.14}$$

Besides, the feedback coefficient G_{ad_i2} can also be an SOGI with a negative gain, in which case the GCF resembles a virtual impedance Z_{VI} in series with the L_2, and the expression of Z_{VI} can be derived as

$$Z_{VI}\left(s\right) = \frac{G_{ad\_i2}\left(s\right) G_d\left(s\right) K_{PWM}}{s^2 L_1 C} \tag{2.15}$$

where the SOGI with the phase lead characteristics is utilized to realize the approximate derivative operation.

Table 2.6 summarizes a comprehensive comparison of the merits and drawbacks of different state-feedback-based damping methods. Figure 2.13 shows the placements of virtual impedances in the *LCL*-filter branches. Referring to the damping effect and the resistor placements of PD methods, it can be concluded that the state-feedback-based damping methods based on Z_{VI3} are the most effective owing to the consistent topology structure with that of PD-4.

Among these damping methods based on Z_{VI3}, the derivative feedback of the capacitor voltage has the advantage of lowest hardware cost since the voltage can be measured for both synchronization operations with the grid and resonance damping, without extra sensor. However, its effectiveness would be easily affected by the high-frequency noise and voltage distortion at PCC, and the damping performance is slightly degraded since the approximate derivation is usually used instead of the direct derivative feedback. Although the number of sensors is minimum for the first-order HPF feedback of current i_2, the effectiveness would be compromised by the effect of amplified high-frequency noise.

In comparison, the proportional CCF is widely applied among the various feedback coefficients due to the merits of simple implementation and sufficient damping performance. Moreover, the single-loop ICF and GCF mentioned in Part B can also be regarded as the virtual impedances in series with L_1 and parallel with C, respectively. It is worth noting that the *LCL*-filter resonance peak can be damped only under the positive virtual resistor conditions, no matter the filter- or state-feedback-based damping methods. Nevertheless, the digital control delay caused by algorithm execution has a significant influence on the characteristics of virtual resistor.

Table 2.6 Comparison of state-feedback-based damping methods [5, 23, 26–28]

Feedback signals	Feedback coefficients	Merits	Drawbacks	Major technologies
i_C (capacitor current)	Proportional feedback	• Simple implementation • Superior damping performance	• Impair the phase margin of system • Degraded transient response	• Select G_{ad_ic} with a tradeoff between the gain margin and phase margin
	Proportional integral feedback	• Simple implementation • Improved robustness	• Integral term would accumulate the noise	• Replaced by proportional feedback of i_C plus proportional feedback of u_C
	First-order HPF	• High robustness • Mitigate phase lag	• Amplify high-frequency noise • Cause phase error • Reduce system bandwidth	• Select the proper cutoff frequency of HPF firstly
	Second-order HPF	• Stabilize current loop • Strengthen interaction stability between the inverter and the grid	• Increase steady-state error • Decrease control bandwidth • Amplify high-frequency noise	• Improve the system robustness against grid impedance variation
u_C (capacitor voltage)	HPF	• Magnitude-frequency characteristic is very similar to ideal derivative	• Enlarged phase error at high frequency • Amplify high-frequency noise	• Discretized by Tustin method
	Lead-lag network	• Small phase error below the frequency of maximum phase shift	• Sensitive to the variation of f_r • Narrow available frequency range	• Discretized by prewarped Tustin method at f_r • Maintain phase lead $90°$ at f_r
	Nonideal GI	• Wide derivative range • Avoid noise amplification	• Introduce phase error at high frequency • Complex control algorithm	• Discretized by first-order hold method at Nyquist frequency
	Q-SOGI	• Produce accurate phase at f_r • Avoid noise amplification	• Sensitive to the variation of f_r • Narrow available frequency range	• Discretized by prewarped Tustin method at f_r
i_1 (inverter-side current)	Proportional feedback	• Simple implementation	• Deficient damping effect • Low robustness	• Select proportional coefficient deliberately
	Proportional feedback	• Simple implementation	• May excite additional resonance between C and L_2 • Degraded dynamic tracking performance	• Select proper G_{ad_i1} according to the requirements of damping ratio and f_r
	WAC	• Simplify controller design • Improve the closed-loop control bandwidth	• May appear two resonance frequencies • Sensitive to grid impedance variation	• Improve the system robustness against grid impedance variation
i_2 (grid-side current)	First-order HPF	• Simple control algorithm • Low hardware cost	• Amplify high-frequency noise • Reduced system bandwidth	• Select the gain and cutoff frequency of HPF eclectically

Fig. 2.13 Equivalent circuit of state-feedback-based damping methods

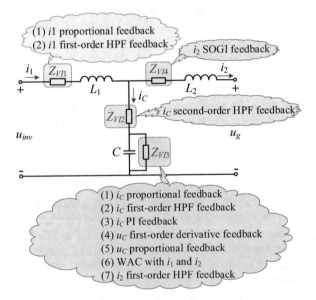

(1) $i1$ proportional feedback
(2) $i1$ first-order HPF feedback

i_2 SOGI feedback

i_C second-order HPF feedback

(1) i_C proportional feedback
(2) i_C first-order HPF feedback
(3) i_C PI feedback
(4) u_C first-order derivative feedback
(5) u_C proportional feedback
(6) WAC with i_1 and i_2
(7) i_2 first-order HPF feedback

2.3.4 Effect of Control Delay and Application Issues

The total digital control delay T_d composed of computation and PWM delay inherently exists in the current control loop, and the typical values of the total control delay in practical situations are $1.5T_{sm}$ and T_{sm}, where T_{sm} is the sampling period. It is well known that the phase margin and control bandwidth of the system are decreased due to the phase lag introduced by control delay, and the lagging phase can be expressed as follows:

$$\varphi_{delay} = -\frac{f_r}{f_{sm}} \times \left(\frac{T_d}{T_{sm}} + 0.5 \right) \times 360° \tag{2.16}$$

where f_{sm} is the sampling frequency and f_r is the LCL-filter resonance frequency.

As for the single-loop ICF, an inherent damping term is embedded in the control loop to stabilize the system in the case of ignoring the control delay, whereas the system stability is deteriorated with the consideration of control delay. Hence, the influence of control delay should be reduced for ICF. Conversely, with respect to the single-loop GCF, the system is unstable without control delay, yet inherently stable owing to the proper control delay, also called the inherent damping. This is why the stability of the system with GCF can be improved by means of phase-lag filters. Note that the system is also unstable if the lagging phase is too large. The inherent damping of ICF and GCF is produced by the positive virtual resistor, whereas the system may be unstable when the virtual resistor is negative.

Figure 2.14 shows the stability regions of the single-loop ICF and GCF without any damping measures when $T_d = 1.5T_{sm}$, where $f_{sm}/6$ is the critical frequency f_c in this case, but the f_c is $f_{sm}/4$ when $T_d = T_{sm}$. According to the NSC, the system is

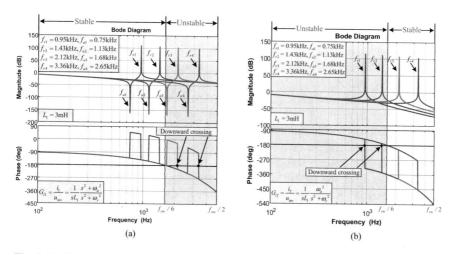

Fig. 2.14 Open-loop Bode diagrams of different current feedback strategies, with the typical value of $T_d = 1.5T_{sm}$ [29]. (**a**) ICF. (**b**) GCF

unstable due to the unequal numbers of upward and downward $-180°$ crossings in the frequency range with the gain above 0 dB. Corresponding to the unstable regions in Fig. 2.14, the virtual impedance involves a negative resistor element[11]. Obviously, in order to widen the stable regions for a positive virtual resistor of a wider range, the *LCL*-filter resonance frequency can be properly shifted, whereas this method is not practically feasible due to the demand of redesigned *LCL*-filter parameters. Other methods are to adjust the sampling frequency or to control the delay correctly.

In Fig. 2.14, the stable region of ICF can be widened by enlarging the f_c through decreasing the delay effect, but the time delay of GCF should be appropriately increased to lower the f_c for broadening the stable region. For the widely adopted proportional CCF, its stable region is consistent with that of ICF in Fig. 2.14a. Similarly, the negative virtual resistor contained in the virtual impedance is disadvantageous to system stability. To solve this problem, the predictive control techniques, filter-based compensation methods, and the modified sampling methods are alternative solutions to diminish the impact of control delay, such as the Smith prediction, the state observer, and the linear prediction. Furthermore, the filter-based methods are also candidates for delay compensation with the phase lead characteristics, such as the SOGI, the lead-lag compensator, HPF, and the infinite impulse response (IIR) filter.

Different from the predictive techniques and filter-based compensation approaches, the modified sampling methods can be more intuitively and easily utilized to reduce delay, which consists of multiple sampling, shifting the sampling instant, shifting the updating instant of reference voltage and dual sampling modes. The control delay can be reduced by multiple sampling method with enlarged sampling frequency; therefore, the stability region of the system is enlarged. However, this method would result in computation burden on the microprocessor and multiple intersections between the modulation wave and the carrier. It is worth noting that the PWM delay ($0.5T_{sm}$) cannot be decreased when the sampling frequency is fixed. In this case, the computation delay can be reduced by shifting the sampling instant toward the update instant of the modulation signal; thus, the total digital control delay is diminished. However, the sampled signals are susceptible to the aliasing and switching noises.

Table 2.7 summarizes a comprehensive comparison to show the merits and drawbacks of various methods to reduce the effect of control delay. Undoubtedly, the modified sampling methods are most straightforward, next are filter-based compensation approaches, and then are the predictive control schemes. However, the influence of delay on system stability can only be alleviated rather than thoroughly eliminated by applying modified sampling methods, except for the dual sampling modes. Therefore, a reasonable tradeoff between the implementation complexity and the compensation effect could be attained by means of digital filters with phase lead characteristics. Certainly, by using the delay compensation approaches and various damping methods, the internal stability of individual inverter itself can be improved significantly. However, the interactions between the inverter and the grid, as well as among paralleled inverters, still challenge the external stability of the

Table 2.7 Merits and drawbacks of the methods for reducing delay influence [23, 29]

	Specific methods	Merits	Drawbacks
Predictive control schemes	Smith prediction	• Compensate delay completely • High control bandwidth	• Sensitive to system parameters • Complicated algorithm
	State observer	• Compensate delay completely • High control bandwidth	• Sensitive to system parameters • Computation burden
	Linear prediction	• Easy implementation • Robust against parameters variation	• Deficient delay compensation at high frequency • Limited control bandwidth
Filter-based compensation methods	Lead-lag compensator	• Simple implementation • Approximate a differentiator at f_c	• Relatively complex parameters design • Sensitive to the variation of f_c • Exist phase error below 90
	SOGI	• Simple implementation • High robustness	• Noise amplification at high frequency • Relatively poor compensation ability
	IIR filter	• Simple implementation • Accurate phase compensation	• Noise amplification at high frequency • Relatively complex parameters design
	HPF	• Simple implementation • Ideal derivative characteristics at low frequency	• Over compensation at low frequency • Noise amplification at high frequency
Modified sampling methods	Multiple sampling	• Reduce delay • Increase the control bandwidth	• Introduce the switching noise and aliasing • Increased computational burden
	Shifting the sampling instant	• Reduce delay • Easy implementation with simple algorithm	• Introduce the switching noises and aliasing • Reduced computational time • Limited by computational speed
	Shifting the update instant of reference voltage	• Reduce delay • Easy implementation with simple algorithm	• Limited by computational speed
	Dual sampling modes	• Eliminate delay completely • Extend the sampling interval • Improve the current control performance • Avoid switching noises	• Complicated algorithm • Computational burden • Unsuitable for three-phase grid-connected inverters

inverter, and therefore the corresponding concerns about the external interaction stability of the inverter-grid system are discussed in the forthcoming section.

2.4 Impedance-Based Method for External Stability

To apply the impedance-based stability criterion, a suitable impedance model of the inverter should be obtained first by employing reasonable modeling methods. Furthermore, grid impedance is supposed to be acquirable through relevant measurement techniques.

2.4.1 Impedance-Based Stability Criterion

Originally utilized to analyze the stability of DC systems, the impedance-based stability analysis method is then introduced to the stability analysis of AC systems. The current-controlled inverter-grid system can be separated as an inverter subsystem and a grid subsystem by applying the impedance-based analysis method, in which case the inverter and the grid can be respectively denoted by a current source in parallel with an impedance and a voltage source in series with an impedance, as shown in Fig. 2.15, where $u_g(s)$ is the grid voltage, $u_{pcc}(s)$ is the voltage at PCC, $i_2(s)$ is the injected grid current, and $i_s(s)$ is the ideal current source [23, 31].

In Fig. 2.15a, the equivalent output impedance of the inverter is defined as $Z_o(s)$, and the grid impedance $Z_g(s)$ is mainly dependent on the transmission lines and transformer impedances. The injected grid current can be expressed by (2.17),

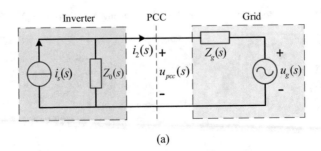

(a)

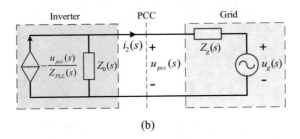

(b)

Fig. 2.15 Equivalent circuit of the inverter-grid system [10, 30]. (**a**) When the PLL is neglected. (**b**) When the PLL is considered

which indicates that a large value of $Z_o(s)$ represents a good steady-state performance.

$$i_2(s) = \left[i_s(s) - \frac{u_g(s)}{Z_o(s)} \right] \frac{1}{1 + Z_g(s)/Z_o(s)} \tag{2.17}$$

According to [10], the interconnected inverter-grid system is stable if one of the following conditions is satisfied, which can be achieved by either cancelling the equivalent grid impedance or increasing the inverter output impedance.

1. $Z_g(s) = 0$.
2. $Z_g(s)/Z_o(s)$ satisfies the Nyquist stability criterion.

Furthermore, the inverter can be regarded as a controlled current source with the effect of phase-locked loop (PLL), as shown in Fig. 2.15b, where $Z_{PLL}(s)$ is the equivalent impedance of PLL. In this scenario, the injected grid current can be calculated by (2.18), and the stability conditions are consistent with those of Fig. 2.15a.

$$i_2(s) = -\frac{u_g(s)}{Z_{inv}(s)} \left(\frac{1}{1 + Z_g(s)/Z_{inv}(s)} \right) \tag{2.18}$$

In (2.18), $Z_{inv}(s)$ is the equivalent output impedance of the inverter, which is the parallel form of $Z_o(s)$ and $Z_{PLL}(s)$.

2.4.2 Impedance Modeling Methods

According to the relationship between the voltage at PCC and the injected grid current, the impedance model based on the equivalent transfer function can be derived by means of direct linearization in steady state. However, the practical AC systems are nonlinear due to switching devices and controllers, which can be approximately linearized by applying the conventional small-signal methods. It is worth noting that the direct small-signal linearization is incapable for time-varying AC systems due to the absence of the fixed steady-state operating point, in which case the variables need to be represented in the dq-, sequence-, or phasor domains, thereby the small-signal linearization can be executed for modeling and stability analysis.

1. Equivalent Transfer Functions

By using the equivalent transformation of the control block diagrams of the system, the inverter output impedance can be calculated. Concretely, the block diagrams of the transfer functions of system can be equivalently transformed as shown in Fig. 2.16.

The $G_1(s)$ and $G_2(s)$ are the equivalent terms of the system block diagrams after transformation, which vary with the different damping methods and controller parameters. The H_{i2} is the sensor gain of grid current, I_{ref}^* is the amplitude command of the current reference i_{ref}, θ is the extracted voltage phase of PCC by the synchronous rotating reference-frame PLL (SRF-PLL), and φ is the power-factor angle. The phase of i_{ref} is $(\theta$-$\varphi)$ since the grid-connected inverter is required to output

Fig. 2.16 The equivalent transfer function block diagrams of the system [12, 30]. (**a**) When the SRF-PLL is neglected. (**b**) When the SRF-PLL is considered

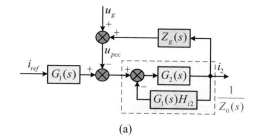

(a)

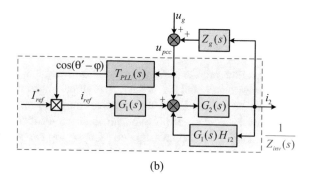

(b)

reactive power. The T_{PLL} is the transfer function model of SRF-PLL. In this scenario, the equivalent output impedances of the inverter in Fig. 2.16 can be respectively derived as follows:

$$Z_o = -\frac{u_{pcc}}{i_2} = \frac{1+G_1G_2H_{i2}}{G_2} \tag{2.19}$$

$$Z_{inv} = -\frac{u_{pcc}}{i_2} = \frac{1+G_1G_2H_{i2}}{G_2 - I^*_{ref}T_{PLL}G_1G_2} \tag{2.20}$$

In (2.20), the equivalent output impedance of the inverter is decreased due to the additional polynomial $-T_{PLL}I^*_{ref}G_1G_2$ introduced by SRF-PLL regardless of the damping methods, which deteriorates system stability. The feedforward scheme of grid voltage can be applied to reshape the output impedance phase of the inverter at the intersection frequency, where the magnitudes of Z_o and Z_g are equal, thereby enlarging the phase margin of the system. However, the system stability is susceptible to the negative phase angle introduced by the feedforward strategy. Fortunately, it can be solved by boosting the phase angle of inverter output impedance with the utilization of an improved feedforward strategy. Apart from the single-phase inverter systems, this method can also be applied to the three-phase systems controlled in the $\alpha\beta$ frame since no coupling effect exists between the α- and β-axis components.

2. DQ-Domain Impedance Modeling

The impedance models in the dq-domain are easily compatible with each other in an overall system model since most of the three-phase systems are controlled in the dq-frame without additional transformation. With respect to a three-phase inverter system controlled in the dq-frame, the three-phase variables are transformed into two DC quantities, that is, the d- and q-axis components. In this case, the equivalent output impedance of the inverter in the dq-domain can be obtained by using the conventional small-signal linearization method around the DC operation points, and each link in the transfer function flowchart is represented by a two-dimensional matrix. Hence, the terminal characteristics of the inverter can be represented by an impedance matrix as follows:

$$Z_{inv} = \begin{bmatrix} \tilde{v}_d \\ \tilde{v}_q \end{bmatrix} \begin{bmatrix} \tilde{i}_d \\ \tilde{i}_q \end{bmatrix}^{-1} = \begin{bmatrix} Z_{dd} & Z_{dq} \\ Z_{qd} & Z_{qq} \end{bmatrix} \tag{2.21}$$

where $\tilde{i}_d$ and $\tilde{i}_q$ are the current perturbation signals, $\tilde{v}_d$ and $\tilde{v}_q$ are the corresponding voltage responses, and Z_{inv} is the output impedance matrix of the inverter.

It can be deduced that the crossing-coupling impedances Z_{dq} and Z_{qd} can be neglected in the case of unity power factor controlled inverters, in which case the system stability can be investigated by applying two single-input and single-output (SISO) models. Moreover, the stability of the inverter system is susceptible to the negative incremental resistance Z_{qq} introduced by PLL at low frequencies, whereas the negative resistance characteristics may exist in Z_{dd} for rectifier systems. In order to modify the q-axis output impedance Z_{qq} into a positive resistance, a feedforward control strategy of the voltage at PCC can be adopted. Yet, the inaccurate stability analysis may be caused by the inherent phase drop of the inverter output impedance, which is introduced by the control delay in the feedforward path, and this problem can be solved by means of reduced feedforward gain.

Nevertheless, several uncertainties in practice, such as the unbalanced three-phase AC systems and the phase tracking error of PLL, may influence the accuracy of stability analysis, which needs to be further explored. Moreover, with respect to the d- and q-axis impedances, there is unspecific physical interpretation due to the artificial frame. The d- and q-axis terminals are not physical existing for connecting sensors, and the dq-frame must be processed in real time, which results in inconvenient measurement owing to the special requirements of measurement algorithms and devices.

3. Phasor-Domain Impedance Modeling

In phasor domain, the sinusoidal voltages and currents can be respectively denoted by the phasors composed of two-state variables for more accurate system modeling, in which case the phasors are DC quantities in steady state. Thus, a 2×2 phasor-domain impedance matrix can be defined by injecting sinusoidal disturbances of real and imaginary parts, which allows the direct linearization of single-phase system for impedance-based stability analysis. Moreover, the terminal characteristics of the inverter can also be described by defining the amplitude perturbation $\tilde{V}_m$ and phase perturbation $\tilde{V}_\theta$ at fundamental frequency, expressed as (2.22).

$$\begin{bmatrix} \tilde{I}_m(s) \\ \tilde{I}_\theta(s) \end{bmatrix} = \begin{bmatrix} Y_{mm}(s) & Y_{m\theta}(s) \\ Y_{\theta m}(s) & Y_{\theta\theta}(s) \end{bmatrix} \begin{bmatrix} \tilde{V}_m(s) \\ \tilde{V}_\theta(s) \end{bmatrix} \tag{2.22}$$

where $\tilde{I}_m$ and $\tilde{I}_\theta$ are the current responses, and the transfer matrix is the inverter output admittance in phasor domain.

It is worth noting that the impedance modeling and stability analysis of the inverter in phasor domain is immature at present, which needs to be further explored.

4. Sequence-Domain Impedance Modeling

Harmonic linearization techniques can be used to derive a linearization model of an inverter along a sinusoidal trajectory. By superimposing the positive- and negative-sequence sinusoidal voltage perturbations to the time-varying operating trajectory, the inverter output impedance in the sequence domain can be obtained according to the current responses at the perturbation frequency, which is denoted by an admittance matrix as follows:

$$\begin{bmatrix} \tilde{I}_p\left(s+j\omega_0\right) \\ \tilde{I}_n\left(s-j\omega_0\right) \end{bmatrix} = \begin{bmatrix} Y_{pp}\left(s\right) & Y_{pn}\left(s\right) \\ Y_{np}\left(s\right) & Y_{nn}\left(s\right) \end{bmatrix} \begin{bmatrix} \tilde{V}_p\left(s+j\omega_0\right) \\ \tilde{V}_n\left(s-j\omega_0\right) \end{bmatrix} \tag{2.23}$$

where the subscripts p and n represent the positive- and negative-sequence components, respectively, $\tilde{V}$ is the voltage perturbation, $\tilde{I}$ is the current response, and ω_0 is the fundamental angular frequency of grid voltage.

In order to analyze the merits and drawbacks of the various impedance modeling methods, a comprehensive comparison is conducted in Table 2.8.

Actually, the equivalent transfer functions are SISO models, in which case the inverter systems are regarded as linear and time-invariant (LTI) systems. Hence, the classical gain margin (GM) and phase margin (PM) in a Bode diagram can be employed to design the reasonable system parameters, thereby reshaping the equivalent inverter output impedances for improving system stability. However, the impedance models in dq-, phasor-, and sequence domains are multiple-input and multiple-output (MIMO) transfer matrixes. In this scenario, system stability should be evaluated by means of generalized Nyquist stability criterion (GNSC), which can be denoted as follows:

$$T\left(s\right) = Z_{inv}\left(s\right)Y_g\left(s\right) \tag{2.24}$$

where $Z_{inv}(s)$ is the impedance matrix of the inverter and $Y_g(s)$ is the grid admittance matrix. The eigenvalues $\lambda_1(s)$ and $\lambda_2(s)$ of $T(s)$ are two characteristic loci that vary with the variable s in the complex plane, and the interconnected system is stable if and only if the $(-1, j0)$ is not encircled by the Nyquist curves of $\lambda_1(s)$ and $\lambda_2(s)$. Nevertheless, the GNSC is usually utilized to assess system stability, and the relevant research work on how to design system parameters according to the assessment results is still insufficient.

Table 2.8 Merits and drawbacks of the different impedance modeling methods [12, 23, 30]

Impedance modeling methods	Specific implementation methods	Merits	Drawbacks	Applications
Equivalent transfer functions	• Transform the transfer function block diagrams equivalently	• Clear physical interpretation for the impedances • Simple implementation • Facilitate parameters design	• Relatively low model accuracy	• Single-phase systems • Balanced three-phase systems controlled in $\alpha\beta$-frame
dq-domain impedance modeling	• Small-signal linearization in dq-frame • Represent the impedance by 2-demensional matrix	• Three-phase inverters are usually controlled in dq-frame • Compatible with overall system model	• Exist coupling between d- and q-axis components • Cannot measure d- and q-axis components through experiments directly	• Balanced three-phase systems
Phasor-domain impedance modeling	• Represent the state variables with phasors	• Describe the system model accurately	• Immature impedance-based stability analysis	• Single-phase systems • Balanced three-phase systems
Sequence-domain impedance modeling	• Harmonic linearization and symmetrical component method	• Clear physical interpretation for the impedance models • Impedances can be measured through experimentations directly	• Exist coupling between sequence impedances • Complicated linearization process	• Three-phase systems

Indeed, the analytical inverter output impedance can be acquired by means of the aforementioned theoretical derivation. Yet, the impedance model is unavailable due to the missing system parameters in some practical cases, which impedes the application of impedance-based stability analysis method. Fortunately, the output impedance can also be measured by capturing the terminal characteristics of the inverter through injecting the perturbation signals into the grid voltage, in which case the system is regarded as a black box. Moreover, the theoretical derivation method is unsuitable for modeling grid impedance and, therefore, the relative measurement techniques should be adopted. Hence, the signal-injection based impedance measurement techniques are discussed herein.

2.4.3   Online Impedance Measurement Techniques

By injecting a sinusoidal excitation into the system at the predetermined frequency and capturing the output response at that frequency, the sine sweep technique is applied to calculate the impedance through point-by-point measurement in a specific frequency range with several frequency points, which is reliable due to its high signal-to-noise ratio (SNR). However, as for the time-varying grid impedance, the off-line measurement is insufficient for accurate stability analysis. In this scenario, the sine sweep technique is not suitable anymore on account of the time-consuming measurement and, therefore, online impedance measurement methods with short processing time can be adopted for the online stability analysis in real time, thereby reshaping the inverter output impedance adaptively. The injection signals could be the single impulse or the pulse sequences, in which case the signal contains abundant harmonics information. Therefore, the grid impedance can be accurately estimated within a period of about one fundamental cycle.

As shown in Fig. 2.17, in these online methods, by injecting a current perturbation on top of the current reference i_{dref} (i_{qref}) and measuring the voltage responses, the corresponding frequency components in response and perturbation are extracted by means of Fourier analysis; then the grid impedance is timely acquired by the ratio between the voltage u_{Cd} (u_{Cq}) and the current i_{2d} (i_{2q}) at different frequencies, without extra sensors and signal generators.

Fig. 2.17 Schematic diagram of the online grid impedance identification based on perturbation signals injection [14]

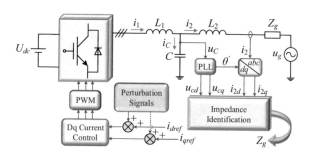

Specifically, the advantages and disadvantages of various measurement techniques are summarized in Table 2.9. In conclusion, the processing time of single impulse injection is shorter than that of the binary- and ternary-sequence injection methods, whereas the latter two approaches are more reliable in view of the system stability due to the smaller time-domain amplitudes. Hence, the pulse sequences are suitable for the perturbed sensitive systems. As for the selection of the perturbation signals, it is worth noting that the signal amplitude should be carefully determined, since a large amplitude may interfere with the normal operation of the system, yet a small amplitude is deficient for impedance measurement due to the low SNR. Another consideration is the generation frequency of the sequences; a high value of frequency is advantageous to provide enough energy for the sequences, yet a low value, below one-third of the sampling frequency of the inverter, contributes to achieving precise measurement.

2.4.4   Stability Analysis of Multi-paralleled Grid-Connected Inverters

Multi-paralleled inverters are increasingly being employed to enlarge the total generation capacity in recent years. In this scenario, apart from the interaction between the grid and the inverter, the additional interaction instability among inverters may also arise as the number of inverters increases.

The Norton equivalent circuits of multiple paralleled inverters in Fig. 2.18a, which imparts that both types of interactions in the parallel system would cause the distorted grid current. Specifically, the interaction among inverters is mainly caused by the change of current references of other inverters, such as in a PV plant where the current references are generally varied in every several seconds, and the interaction between the grid and the inverter is induced by the grid voltage harmonics and

Table 2.9 Advantages and disadvantages of the online impedance measurement techniques [14, 23]

Injected signals	Advantages	Disadvantages	Applications
Single impulse	• Simple implementation • Insensitive to the no-stationary nature of AC systems • Short testing time	• May interfere with the normal operation of the inverter • Relatively low measurement accuracy	• Insensitive systems for perturbations • Weak grid
MLBS	• Small excitation amplitude • Simple implementation • Easy data acquisition	• Neglect the crossing-coupling effect between the impedance components • Cannot measure the impedance components in the same operation conditions • Energy is distributed over many harmonic frequencies	• Sensitive systems for perturbations • Balanced three-phase systems • Weak grid
DIBS	• Small excitation amplitude • Simple implementation • Easy data acquisition • Maximize the energy of the signals	• Neglect the crossing-coupling effect between the impedance components • Cannot measure the impedance components in the same operation conditions	• Balanced three-phase systems • Strong grid
OPRBS	• Save the test time • Measure each impedance component in the same operation conditions	• Neglect the crossing-coupling effect between the impedance components • Sequence lengths are different	• Balanced three-phase systems • Weak grid
Ternary sequence	• Much wider of the sequence length • More efficient • Minimize the effect of nonlinearities	• Neglect the crossing-coupling effect between the impedance components • Cannot measure the impedance components in the same operation conditions	• Balanced three-phase systems • Weak grid

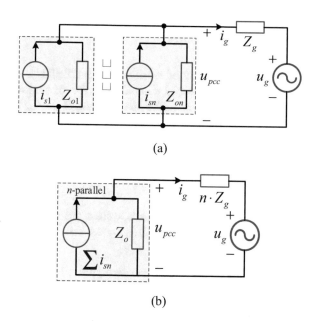

(a)

(b)

Fig. 2.18 (a) Equivalent circuit of n-parallel inverters. (b) Simplified equivalent circuit of n-parallel inverters [32]

transient disturbances. Further, in order to simplify the system model, an equivalent inverter can be used to model n paralleled identical inverters, which reveals the impedance multiplication effect, that is, the equivalent grid impedance is n times the original value, as shown in Fig. 2.18b. In this case, the stability of the inverter-grid system is deteriorated due to the n times grid impedance. Nevertheless, the above analysis is ideally implemented with the hypothesis of identical paralleled inverters, and the more general scenarios should be taken into account.

2.5   Benchmark Systems for Stability Analysis

To simplify the analysis procedure for the case studies on stability evaluations of grid-connected inverters, the control block diagram representation is utilized in the following benchmark systems, including the PR controller and SRF-PI controller-based control with capacitor current feedback active damping schemes.

2.5.1 Stability Evaluation Grid-Connected Inverter Using PR Controller Without Delay

Figure 2.19 shows the block diagram of capacitor current feedback active damping for an LCL-type grid-connected inverter without the effect of control delay, where G_i denotes the PR controller, G_{inv} denotes the inverter gain, $G_{inv} = V_{dc}/V_{tri}$, V_{dc} denotes the dc-link voltage of the inverter, and V_{tri} denotes the amplitude of the PWM carrier signal. The PR controller is represented as:

$$G_i\left(s\right) = K_p + \frac{2 * K_r * \omega_i * s}{s^2 + 2 * \omega_i * s + \omega_0^2} \qquad (2.25)$$

The system parameters for the benchmark system are listed as V_g = 220 V, V_{dc} = 360 V, L_1 = 600 μH, L_2 = 150 μH, C = 10 μF, f_r = 4.6 kHz, V_{tri} = 3 V, f_{sw} = 10 kHz, P_o = 6 kW, H_{i2} = 0.15, K_p = 0.45, K_i = 350.

- *The PR controller is simplified as a proportional gain.*

Figure 2.20 shows the inverter output admittance versus grid admittance using simplified PR controller without the effect of control delay when H_{i1} varies from 0 to 0.35. It can be observed that with the increase in the active damping coefficient, the non-passivity region of the inverter output admittance is decreased. Besides, the reduction in grid inductance would result in an unstable system. Moreover, with an increase in the feedback gain H_{i1} from 0.35 to 0.6, the inverter output admittance is shown in Fig. 2.21, it can be observed that the inverter output admittance is passive in the whole frequency range.

It should be noted that the inner current loop stability should be ensured before applying the passivity-based stability analysis. For instance, there would be no overlapped parameter range for the active damping coefficient if the inner current loop stability and admittance passivity are analyzed respectively. Compared with the phase margin (PM)- and gain margin (GM)-based stability analysis approach, the passivity-based stability criterion is more conservative. In this case, the

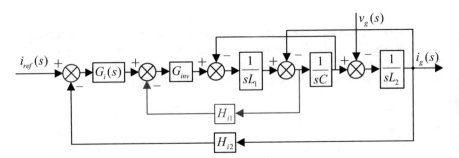

Fig. 2.19 Block diagram of capacitor current feedback active damping for LCL-type grid-connected inverter without the effect of control delay

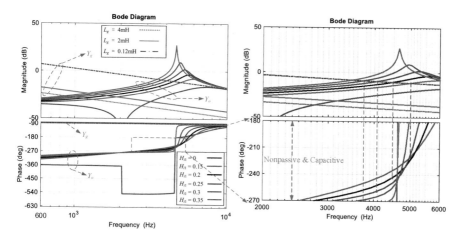

Fig. 2.20 Inverter output admittance versus grid admittance using simplified PR controller without effect of control delay when H_{i1} varies from 0 to 0.35

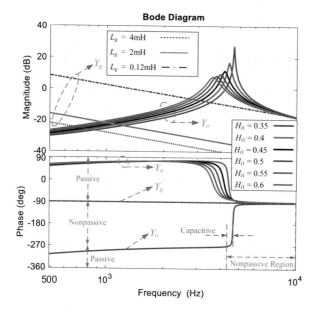

Fig. 2.21 Inverter output admittance versus grid admittance using simplified PR controller without effect of control delay when H_{i1} varies from 0.35 to 0.6

admittance passivity would be compromised for the practical application, to guarantee the inner current loop stability constraints. Normally, the crossover frequency between the inverter output admittance and grid admittance should be outside of the non-passivity capacitive region to ensure system stability.

- *The PR controller is analyzed without simplification.*

Figure 2.22 shows the inverter output admittance versus grid admittance using the PR controller without the effect of control delay when H_{i1} varies from 0 to 0.35. Moreover, with an increase of the feedback gain H_{i1} from 0.35 to 0.6, the inverter output admittance is shown in Fig. 2.23. It can be observed from Figs. 2.22 and 2.23 that with the increase of active damping coefficient, the non-passivity region of the inverter output admittance is increased. Besides, the increase of the grid inductance would result in an unstable system. Therefore, the conclusion deduced from Figs. 2.20 and 2.21 is inconsistent with the case obtained from Figs. 2.22 and 2.23. Hence, the parameter selection criterion is not trivial for the practical system when the impedance-based stability analysis approach is utilized.

Figure 2.24 shows the simulated grid current and PCC voltage with an active damping coefficient $H_{i1} = 0.01$. In this case, the grid current distortion is quite small; however, the PCC voltage is highly distorted, with a THD of 7.08%. Hence, the analytical and simulation results reveal that the parameter range obtained from the simulation verification is smaller than the theoretical analysis due to the unmodeled dynamics of the grid-connected inverter system.

Figure 2.25 shows the simulation results of grid-connected inverter without the effect of control delay, obtained from EMTP-ATP software, where the effect of PR controller model simplification is tested under strong and weak grid conditions. The reference current, the grid-side current, and the tracking error are plotted within the same figure to show the tracking performance, and the signals are denoted in per unit (p.u.) representation. In this case study, the total inductance of the grid-side inductor of LCL and the grid impedance are assumed to be 0.27 mH and 2.7 mH, in case of strong and weak grid scenarios. Other parameters are consistent with the theoretical analysis. It can be observed that in case of strong grid conditions, stable operation of the system is ensured using a simplified and detailed PR controller. However, the simplified PR controller model would result in a higher steady-state error for the current tracking controller. Whereas in case of weak grid conditions, the original detailed PR controller model would result in an unstable system due to a comparatively large controller gain. The simplified PR controller model results in a stable system, with a much higher steady-state error.

2.5.2 Stability Evaluation Grid-Connected Inverter Using PR Controller with Control Delay (Fig. 2.26)

- *The PR controller is simplified as a proportional gain*

    Next, let us consider the effect of control delay, which is denoted as:

$$G_d(s) = e^{-s1.5T_{sm}} \tag{2.26}$$

where T_{sm} represents the sampling period.

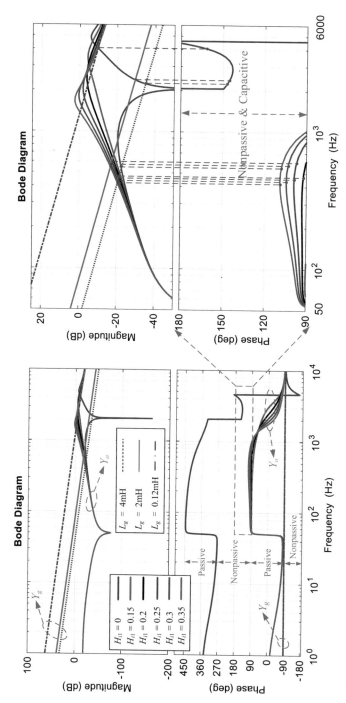

Fig. 2.22 Inverter output admittance versus grid admittance using PR controller without effect of control delay when H_{i1} varies from 0 to 0.35

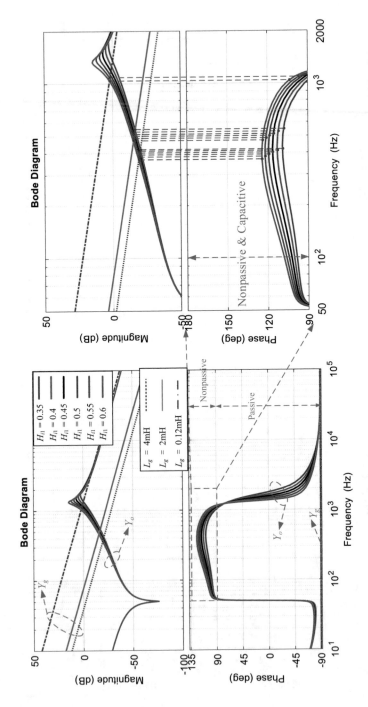

Fig. 2.23 Inverter output admittance versus grid admittance using PR controller without effect of control delay when H_{i1} varies from 0.35 to 0.6

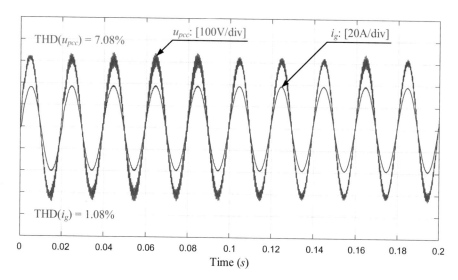

Fig. 2.24 The simulated grid current and PCC voltage with an active damping coefficient $H_{i1} = 0.01$

The system parameters for this case are listed as: $V_g = 220$ V, $V_{dc} = 360$ V, $L_1 = 600\,\mu$H, $L_2 = 150\,\mu$H, $C = 10\,\mu$F, $f_r = 4.6$ kHz, $V_{tri} = 3$ V, $f_{sw} = 10$ kHz, $P_o = 6$ kW, $H_{i2} = 0.15$, $K_p = 0.45$, $K_i = 350$, $K_r = 63$.

Figure 2.27 shows the inverter output admittance versus grid admittance using the simplified PR controller with the effect of control delay when H_{i1} varies from 0 to 0.05. Moreover, with an increase in the feedback gain H_{i1} from 0.05 to 0.1, the inverter output admittance is shown in Fig. 2.28. It can be observed from Figs. 2.27 and 2.28 that, with the increase of active damping coefficient, the non-passivity region of the inverter output admittance is decreased. Besides, the increase of the grid inductance would result in an unstable system.

- *The PR controller is analyzed without simplification*

Figure 2.29 shows the inverter output admittance versus grid admittance using the PR controller with the effect of control delay when H_{i1} varies from 0 to 0.05. Moreover, with an increase of the feedback gain H_{i1} from 0.05 to 0.1, the inverter output admittance is shown in Fig. 2.30. It can be observed from Figs. 2.29 and 2.30 that, with the increase of active damping coefficient, the non-passivity region of the inverter output admittance is increased. Besides, the increase of the grid inductance would result in an unstable system.

Figure 2.31 shows the simulation results of grid-connected inverter with the effect of control delay and the simplified PR controller model, where the effective control delay $T_d = 1/2T_s$, T_s and $3/2T_s$ are considered. Besides, the effects of strong and weak grid are compared under different control delay scenarios. It can be observed that the weak grid condition is more immune to the effect control delay, that is, the system stability is guaranteed under different control delay conditions.

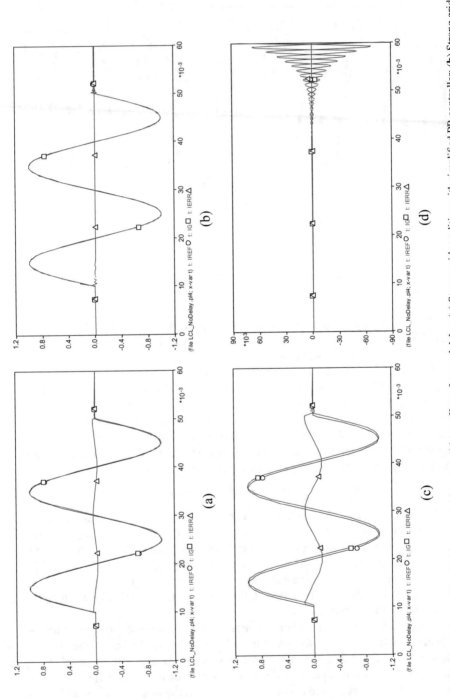

Fig. 2.25 Simulation results of grid-connected inverter without effect of control delay. (**a**) Strong grid condition with simplified PR controller; (**b**) Strong grid condition with detailed PR controller; (**c**) Weak grid condition with detailed PR controller; (**d**) Weak grid condition with detailed PR controller

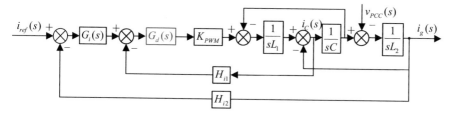

Fig. 2.26 Block diagram of capacitor current feedback active damping for LCL-type grid-connected inverter with the effect of control delay

Whereas in case of a strong-grid scenario, a larger control delay would result in an unstable system, as illustrated in the simulation waveforms in Fig. 2.31.

Figure 2.32 shows the simulation results of grid-connected inverter with the effect of control delay and the detailed PR controller model, where the effective control delay $T_d = 1/2T_s$, T_s and $3/2T_s$ are considered. Besides, the effects of strong and weak grid are also compared under different control delay scenarios. Similar results have been obtained, which are consistent with the case in Fig. 2.31. However, in case of the detailed PR controller model, the current tracking error is remarkably reduced in case of the stable system. And a little transient oscillation can be observed when the detailed PR controlled is applied due to the sluggish transient response, as indicated in the existing literatures.

Figure 2.33 shows the simulation results of the grid-connected inverter with the effect of control delay and detailed PR controller model under weak grid condition when $T_d = 3/2T_s$. It shows that with a decrease of the resonant gain of the PR controller, the system can be stabilized with less transient overshoot and enhanced dynamic performance.

Figure 2.34 shows the case with the effect of control delay and detailed PR controller under weak grid condition, when $T_d = 3/2T_s$ with a decrease in the active damping coefficient. It shows that with a decrease of active damping coefficient, the system can be stabilized with less transient oscillations and improved dynamic tracking performance.

The simulation results obtained from EMTP-ATP indicates that the weak grid condition is more immune to the effect of control delay in terms of closed-loop system stability. Moreover, the regulator gain of the PR controller should be reduced in case of larger control delay, and the active damping coefficient should be decreased in order to ensure global stability and improve transient behaviors of the current tracking controller.

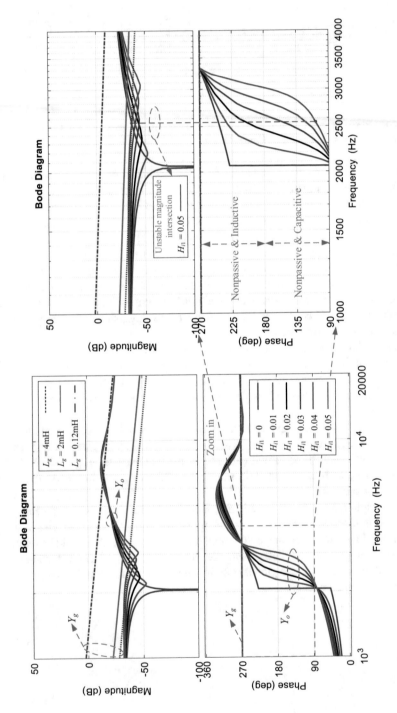

Fig. 2.27 Inverter output admittance versus grid admittance using simplified PR controller with the effect of control delay when H_{i1} varies from 0 to 0.05

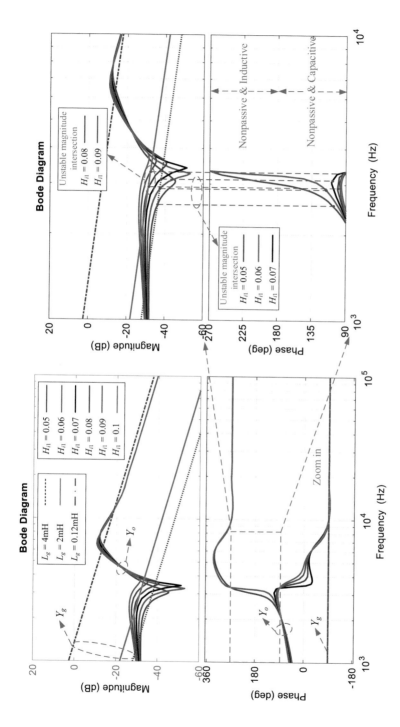

Fig. 2.28 Inverter output admittance versus grid admittance using simplified PR controller with the effect of control delay when H_{i1} varies from 0.05 to 0.1

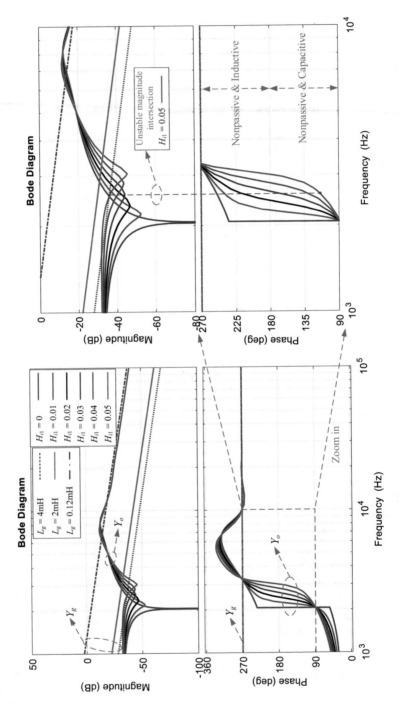

Fig. 2.29 Inverter output admittance versus grid admittance using PR controller with the effect of control delay when H_{i1} varies from 0 to 0.05

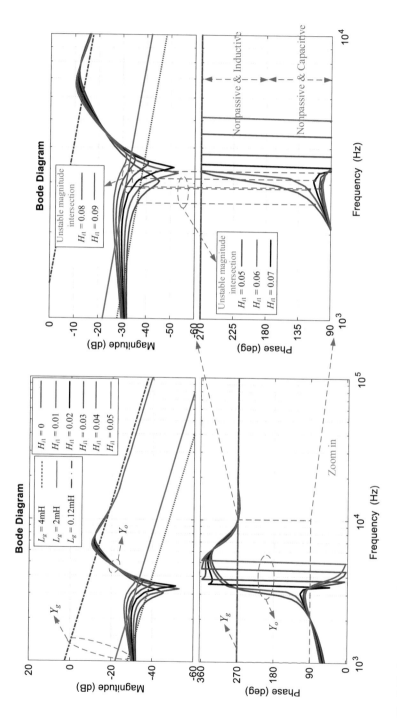

Fig. 2.30 Inverter output admittance versus grid admittance using PR controller with the effect of control delay when H_{i1} varies from 0.05 to 0.1

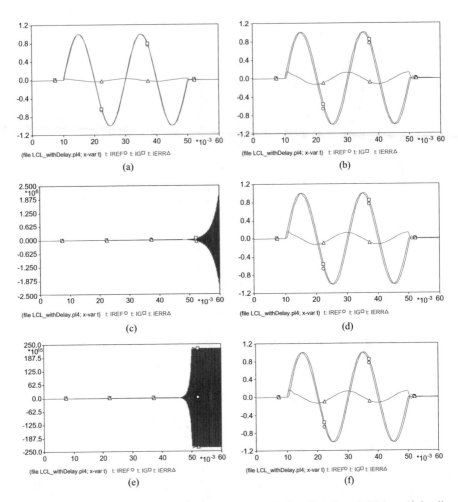

Fig. 2.31 Simulation results of grid-connected inverter with the effect of control delay and simplified PR controller model. (**a**) Strong grid condition with control delay $T_d = 1/2T_s$; (**b**)Weak grid condition with control delay $T_d = 1/2T_s$; (**c**) Strong grid condition with control delay $T_d = T_s$; (**d**) Weak grid condition with control delay $T_d = T_s$; (**e**) Strong grid condition with control delay $T_d = 3/2T_s$; (**f**) Weak grid condition with control delay $T_d = 3/2T_s$

2.5.3 Stability Evaluation Grid-Connected Inverter Using SRF-PI Controller with Control Delay

Next, the current loop controller is replaced using the synchronous frame PI controller, which is denoted as:

$$G_i(s) = \frac{a_3 s^3 + a_2 s^2 + a_1 s + a_0}{s^3 + \omega_0 s^2 + \omega_0^2 s + \omega_0^3} \tag{2.27}$$

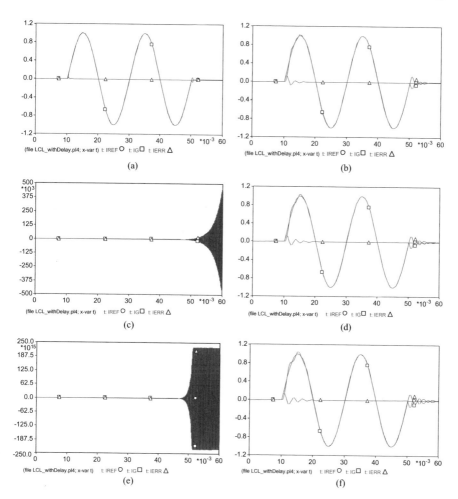

Fig. 2.32 Simulation results of grid-connected inverter with the effect of control delay and detailed PR controller model. (**a**) Strong grid condition with control delay $T_d = 1/2T_s$; (**b**) Weak grid condition with control delay $T_d = 1/2T_s$; (**c**) Strong grid condition with control delay $T_d = T_s$; (**d**) Weak grid condition with control delay $T_d = T_s$; (**e**) Strong grid condition with control delay $T_d = 3/2T_s$; (**f**) Weak grid condition with control delay $T_d = 3/2T_s$

where $a_3 = K_p$, $a_2 = K_p\omega_0 + K_i$, $a_1 = K_p\omega_0^2 + 2\omega_0 K_i$, $a_0 = K_p\omega_0^3 - K_i\omega_0^2$. K_p and K_i denotes the proportional and integral coefficient of the PI controller. And $K_p = 0.35$, $K_i = K_r = 63$ are assumed in the following analysis.

Figure 2.35 shows the inverter output admittance versus grid admittance using the SRF-PI controller with the effect of control delay when H_{i1} varies from 0 to 0.05. Moreover, with an increase of the feedback gain H_{i1} from 0.05 to 0.1, the inverter output admittance is shown in Fig. 2.36. It can be observed from Figs. 2.35 and 2.36 that with the increase of active damping coefficient, the non-passivity region of the

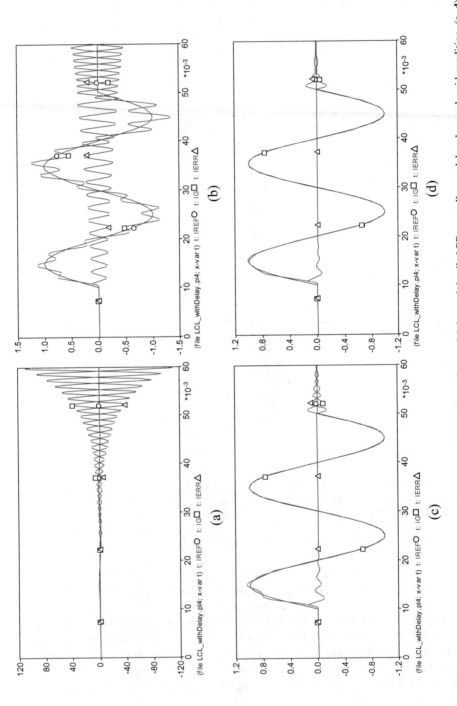

Fig. 2.33 Simulation results of grid-connected inverter with the effect of control delay and detailed PR controller model under weak grid condition. (**a–d**) denotes the case when $T_d = 3/2T_s$ with a decrease of the resonant gain of the PR controller, from 0.6, 0.5, 0.4, to 0.3 pu

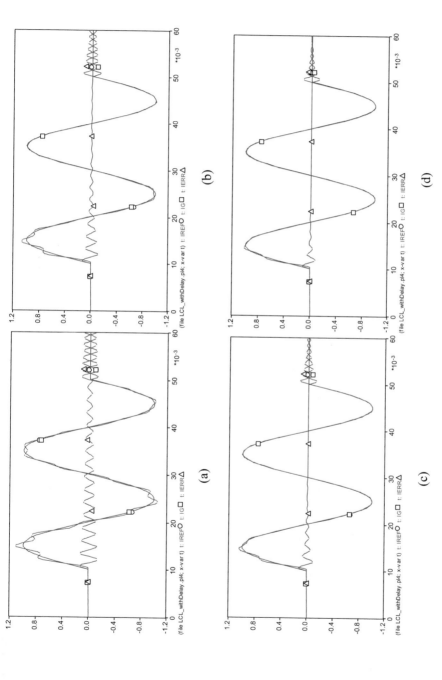

Fig. 2.34 Simulation results of grid-connected inverter with the effect of control delay and detailed PR controller model under weak grid condition. (**a–d**) denotes the case when $T_d = 3/2T_s$ with a decrease of the active damping coefficient, from 0.3, 0.25, 0.2, to 0.15 pu

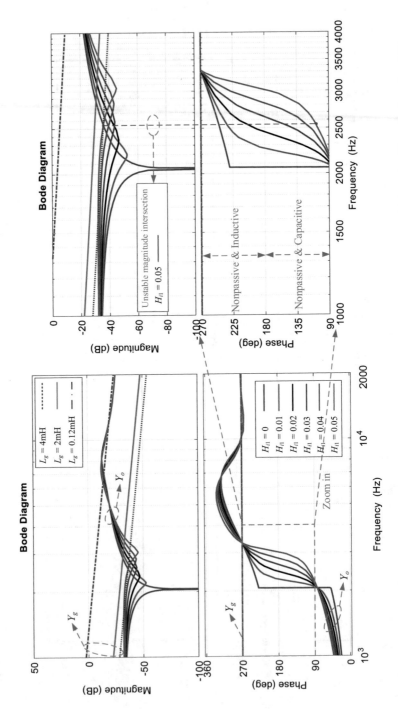

Fig. 2.35 Inverter output admittance versus grid admittance using SRF-PI controller with the effect of control delay when H_{i1} varies from 0 to 0.05

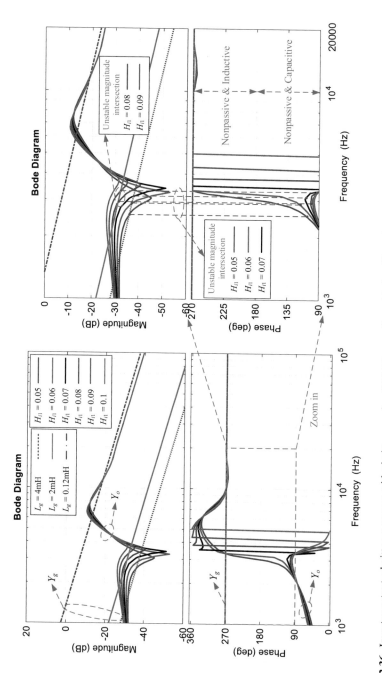

Fig. 2.36 Inverter output admittance versus grid admittance using SRF-PI controller with the effect of control delay when H_{i1} varies from 0.05 to 0.1

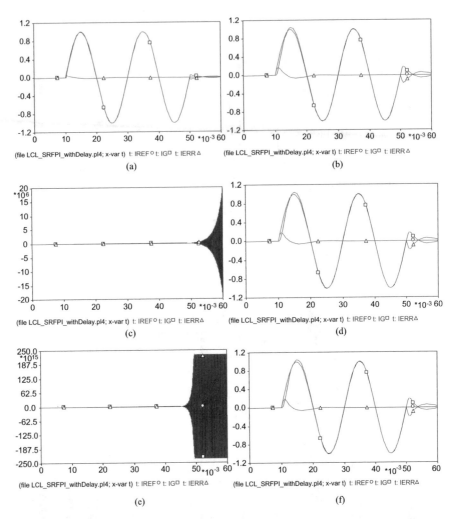

Fig. 2.37 Simulation results of grid-connected inverter with the effect of control delay and SRF-PI controller model. (**a**) Strong grid condition with control delay $T_d = 1/2T_s$; (**b**) Weak grid condition with control delay $T_d = 1/2T_s$; (**c**) Strong grid condition with control delay $T_d = T_s$; (**d**) Weak grid condition with control delay $T_d = T_s$; (**e**) Strong grid condition with control delay $T_d = 3/2T_s$; (**f**) Weak grid condition with control delay $T_d = 3/2T_s$

inverter output admittance is decreased. Besides, the increase in the grid inductance would result in an unstable system.

Figure 2.37 shows the simulation results of the grid-connected inverter with the effect of control delay and the SRF-PI controller model, where the effective control delay $T_d = 1/2T_s$, T_s and $3/2T_s$ are considered. Besides, the effects of strong and weak grid are also compared under different control delay scenarios. In case of strong grid conditions, system stability is more sensitive to the effect of control delay. On the other hand, higher stability margin can be achieved in case of weak grid scenario if identical control parameters are utilized. Figure 2.38 shows the comparison under weak grid conditions with the variation of the active damping coefficients. Instability would occur when the active damping coefficient is not properly selected.

It can be concluded from the previous analysis, without considering the effect of control delay, the results obtained from theoretical analysis would be insistence with the time domain simulation. Besides, due to the equivalent frequency response properties of the PR and SRF-PI controllers, identical results can be obtained in terms of the non-passivity region and stability boundaries with the variation of the active damping coefficient. However, due to the effect of the unmodeled dynamics, the exact consistence in terms of parameter boundaries between the theoretical analysis with the digital simulation or experiment results might be compromised. Nevertheless, the analytical results would provide useful guidelines for the parameter design of the practical grid-connected inverter systems.

2.6 Conclusions

This chapter presents a tutorial of the state-of-the-art techniques in *LCL*-type grid-connected inverters, including the *LCL*-filter parameter design, the damping methods for improving the internal stability of individual inverters, and the impedance-based analysis method for assessing the system-level external stability of inverter-grid systems. The parameters of the *LCL* filter should be meticulously selected according to the design constraints to achieve the desired filtering performance, so that it improves the quality of the injected grid current for avoiding the grid oscillation or even the destabilization caused by harmonics pollution. The specific parameters to be designed include the filter capacitor, the total inductance, the inverter-side inductance, the harmonic attenuation rate, and the resonance frequency. Further, by applying the magnetic integration techniques, the size and weight of the bulky inductors in conventional filters can be diminished to increase the power density of the system.

To maintain the internal stability of *LCL*-type grid-connected inverters, the filter- and state-feedback-based damping methods are preferred to suppress the inherent *LCL*-filter resonance, with the advantages of flexibility, efficiency, and zero power loss. In comparison, the filter-based damping methods are cost-effective and simple to implement, whereas the state-feedback-based methods are more robust in the case of grid impedance variation. However, the digital control delay has significant effects on the two damping methods. As for the external stability at the system level, the impedance-based analysis method is prevalent for predicting the external instability induced by the interactive resonances between the inverter and weak grid, as well as among inverters, especially in the multi-paralleled inverter system. The SISO equivalent transfer functions are suitable for modeling the single-phase and balanced three-phase system controlled in stationary frame. On the other hand, the MIMO transfer matrixes can be utilized to describe the inverter output impedances

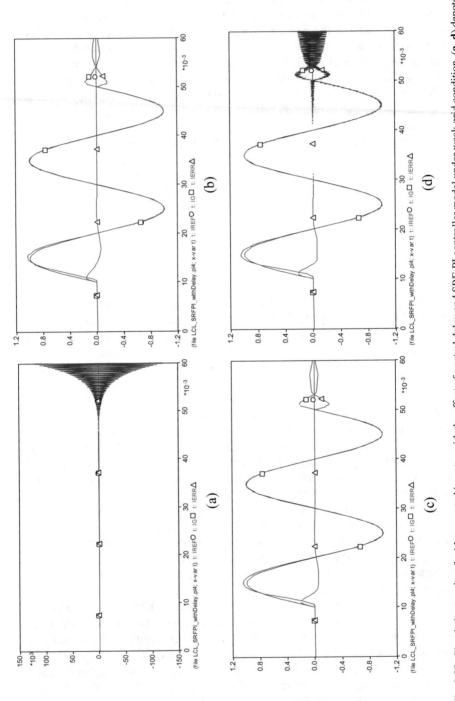

Fig. 2.38 Simulation results of grid-connected inverter with the effect of control delay and SRF-PI controller model under weak grid condition. (**a**–**d**) denotes the case when $T_d = 3/2T_s$ with an increase of the active damping coefficient, from 0.05, 0.1, 0.5, to 0.56 pu

in dq-, phasor-, and sequence domains, in which case the GNSC should be applied to study system stability.

The LCL-type PWM inverter is generally regarded as a linear and balanced system to model its output impedance, whereas the unbalanced characteristics and the several nonlinear factors, such as dead time and PLL, may result in imprecise stability analysis. Furthermore, the lack of simple and applicable stability indexes, similar to the gain and phase margins, impedes the further application of impedance-base stability analysis for systematic control design and parameters tuning. Most of the existing online grid impedance measurement techniques are insufficient, since only the balanced three-phase systems are normally considered and the crossing-coupling effect between the non-diagonal components in the impedance matrix are neglected. Notably, the essence of the online measurement techniques is to determine the impedance values according to the ratios between perturbation signals and the resulting responses and, therefore, the effect of these small disturbances on the system stability should be evaluated in detail.

Regarding the impedance-based stability analysis approach, the contribution of individual inverter to overall system stability in a multi-paralleled inverter application and the identification of the dominant unstable subsystems should be further explored. Furthermore, according to the stability contribution of each inverter, the reasonable impedance sharing control among paralleled inverters is worth considering, to maintain the interaction stability among inverters, as well as the grid-inverter system. In addition, the practical multi-paralleled inverters are not completely identical and, therefore, the more general scenarios containing different types of inverters need to be further studied, such as the differences in power ratings, distribution line impedances, and various types of inverter topologies.

References

1. Blaabjerg, F., Teodorescu, R., Liserre, M., & Timbus, A. V. (2006). Overview of control and grid synchronization for distributed power generation systems. *IEEE Transactions on Industrial Electronics, 53*(5), 1398–1409.
2. Wu, W., Liu, Y., He, Y., Chung, H. S.-H., Liserre, M., & Blaabjerg, F. (2017). Damping methods for resonances caused by *LCL*-filter-based current-controlled grid-tied power inverters: An overview. *IEEE Transactions on Industrial Electronics, 64*(9), 7402–7413.
3. Li, Y. W. (2009). Control and resonance damping of voltage-source and current-source converters with *LC* filters. *IEEE Transactions on Industrial Electronics, 56*(5), 1511–1521.
4. Dannehl, J., Wessels, C., & Fuchs, F. W. (2009). Limitations of voltage-oriented PI current control of grid-connected PWM rectifiers with *LCL* filters. *IEEE Transactions on Industrial Electronics, 56*(2), 380–388.
5. Bao, C., Ruan, X., Wang, X., Li, W., Pan, D., & Weng, K. (2014). Step-by-step controller design for *LCL*-type grid-connected inverter with capacitor–current-feedback active-damping. *IEEE Transactions on Power Electronics, 29*(3), 1239–1253.
6. Liserre, M., Blaabjerg, F., & Hansen, S. (2005). Design and control of an *LCL*-filter-based three-phase active rectifier. *IEEE Transactions on Industry Applications, 41*(5), 1281–1291.

7. Li, X., Lin, P., & Tang, Y. (2018). Magnetic integration of *LTL* filter with two *LC*-traps for grid-connected power converters. *IEEE Journal of Emerging and Selected Topics in Power Electronics, 6*(3), 1434–1446.
8. Ruan, X., Wang, X., Pan, D., Yang, D., Li, W., & Bao, C. (2017). *Control techniques for LCL-type grid-connected inverters.* Springer.
9. Dannehl, J., Liserre, M., & Fuchs, F. W. (2011). Filter-based active damping of voltage source converters with *LCL* filter. *IEEE Transactions on Industrial Electronics, 58*(8), 3623–3633.
10. Sun, J. (2009). Small-signal methods for AC distributed power systems–a review. *IEEE Transactions on Power Electronics, 24*(11), 2545–2554.
11. Wang, X., Harnefors, L., & Blaabjerg, F. (2018). Unified impedance model of grid-connected voltage-source converters. *IEEE Transactions on Power Electronics, 33*(2), 1775–1787.
12. Wang, X., Ruan, X., Liu, S., & Tse, C. K. (2010). Full feedforward of grid voltage for grid-connected inverter with *LCL* filter to suppress current distortion due to grid voltage harmonics. *IEEE Transactions on Power Electronics, 25*(12), 3119–3127.
13. Harnefors, L., Bongiorno, M., & Lundberg, S. (2007). Input-admittance calculation and shaping for controlled voltage-source converters. *IEEE Transactions on Industrial Electronics, 54*(6), 3323–3334.
14. Céspedes, M., & Sun, J. (2012). Online grid impedance identification for adaptive control of grid-connected inverters. In *Proc. IEEE Energy Convers. Congr. Expo. (ECCE)* (pp. 914–921).
15. Wang, T. C. Y., Ye, Z., Sinha, G., & Yuan, X. (2003). Output filter design for a grid-interconnected three-phase inverter. In *Proc. IEEE Aunu. Power Electron. Spec. Conf. (PESC)* (pp. 779–784).
16. Liserre, M., Blaabjerg, F., & Dell'Aquila, A. (2004). Step-by-step design procedure for a grid-connected three-phase PWM voltage source converter. *International Journal of Electronics, 91*(8), 445–460.
17. Fang, J., Li, H., & Tang, Y. (2017). A magnetic integrated *LLCL* filter for grid-connected voltage-source converters. *IEEE Transactions on Power Electronics, 32*(3), 1725–1730.
18. Pan, D., Ruan, X., Bao, C., Li, W., & Wang, X. (2014). Magnetic integration of the *LCL* filter in grid-connected inverters. *IEEE Transactions on Power Electronics, 29*(4), 1573–1578.
19. Lee, K.-J., Park, N.-J., Kim, R.-Y., Ha, D.-H., & Hyun, D.-S. (2008). Design of an *LCL* filter employing a symmetric geometry and its control in grid-connected inverter applications. In *Proc. IEEE Power Electron. Spec. Conf. (PESC)* (pp. 963–966).
20. Li, X., Fang, J., Lin, P., & Tang, Y. (2018). Active magnetic decoupling for improving the performance of integrated *LCL*-filters in grid-connected converters. *IEEE Transactions on Industrial Electronics, 65*(2), 1367–1376.
21. Guo, X.-Q., Wu, W.-Y., & Gu, H.-R. (2010). Modeling and simulation of direct output current control for *LCL*-interfaced grid-connected inverters with parallel passive damping. *Simulation Modelling Practice and Theory, 18*(7), 946–956.
22. Peña-Alzola, R., Liserre, M., Blaabjerg, F., Sebastián, R., Dannehl, J., & Fuchs, F. W. (2013). Analysis of the passive damping losses in *LCL*-filter-based grid converters. *IEEE Transactions on Power Electronics, 28*(6), 2642–2646.
23. Han, Y., Yang, M., Li, H., Xu, L., Coelho, E. A. A., & Guerrero, J. M. (2019). Modeling and stability analysis of LCL-type grid-connected inverters: A comprehensive overview. *IEEE Access, 7*, 114975–115000.
24. Yao, W., Yang, Y., Zhang, X., Blaabjerg, F., & Loh, P. C. (2017). Design and analysis of robust active damping for *LCL* filters using digital notch filters. *IEEE Transactions on Power Electronics, 32*(3), 2360–2375.
25. Pan, D., Ruan, X., & Wang, X. (2018). Direct realization of digital differentiators in discrete domain for active damping of *LCL*-type grid-connected inverter. *IEEE Transactions on Power Electronics, 33*(10), 8461–8473.
26. Shen, G., Zhu, X., Zhang, J., & Xu, D. (2010). A new feedback method for PR current control of *LCL*-filter-based grid-connected inverter. *IEEE Transactions on Industrial Electronics, 57*(6), 2033–2041.

27. Pan, D., Ruan, X., Wang, X., Yu, H., & Xing, Z. (2017). Analysis and design of current control schemes for *LCL*-type grid-connected inverter based on a general mathematical model. *IEEE Transactions on Power Electronics, 32*(6), 4395–4410.
28. Xin, Z., Loh, P. C., Wang, X., Blaabjerg, F., & Tang, Y. (2016). Highly accurate derivatives for *LCL*-filtered grid converter with capacitor voltage active damping. *IEEE Transactions on Power Electronics, 31*(5), 3612–3625.
29. Xin, Z., Wang, X., Loh, P. C., & Blaabjerg, F. (2017). Grid-current-feedback control for *LCL*-filtered grid converters with enhanced stability. *IEEE Transactions on Power Electronics, 32*(4), 3216–3228.
30. Chen, X., Zhang, Y., Wang, S., Chen, J., & Gong, C. (2017). Impedance-phased dynamic control method for grid-connected inverters in a weak grid. *IEEE Transactions on Power Electronics, 32*(1), 274–283.
31. Han, Y., Ning, X., Yang, P., & Xu, L. (2019). Review of power sharing, voltage restoration and stabilization techniques in hierarchical controlled DC microgrids. *IEEE Access, 7*, 149202–149223.
32. Qian, Q., Xie, S., Huang, L., Xu, J., Zhang, Z., & Zhang, B. (2017). Harmonic suppression and stability enhancement for parallel multiple grid-connected inverters based on passive inverter output impedance. *IEEE Transactions on Industrial Electronics, 64*(9), 7587–7598.

Chapter 3
Controller Synthesis and Parameter Selection for Standalone Single-Phase PWM Inverters

Distributed generation (DG) systems, such as photovoltaic and wind power systems, are attracting increasing interest due to the high requirement of reliability and low loss of transmission and distribution networks in recent years. In this chapter, the dual-loop control strategy in hybrid reference frame (HRF) for single-phase voltage source inverters (VSIs) in islanded operation mode is analyzed, which applies a capacitor voltage shaping loop in the synchronous reference frame (SRF) and a capacitor current shaping loop in the stationary reference frame (HRF-based $v + i_c$ control strategy). This control scheme is able to achieve the purpose of active damping, fast dynamic response, and the zero reference tracking error. However, due to the inherent characteristics of SRF-based voltage loop and the digital control delay, the performance of the system is degraded and the control parameter design of HRF-based $v + i_c$ control strategy shows great difficulties. To overcome these shortcomings, in this chapter, a systematic parameter design guideline for HRF-based $v + i_c$ control strategy is proposed to ensure the system stability and optimize the performance of the system under control delay condition. The mathematic model of the HRF-based $v + i_c$ control strategy is established with the consideration of control delay. Based on this model, a satisfactory region of the system stability indexes can be obtained by stability specifications of the system, and the optimal control parameters can be calculated according to the stability indexes selected from the satisfactory region. By using this method, the system stability and robustness can be guaranteed. The simulation results obtained from EMTP and the experimental results obtained from a reduced-scale prototype system are presented to validate the effectiveness of the optimal control parameter design methodologies, which can be widely applied for the similar standalone inverters and uninterruptible power supply (UPS) systems.

© Springer Nature Switzerland AG 2022
Y. Han, *Modeling and Control of Power Electronic Converters for Microgrid Applications*, https://doi.org/10.1007/978-3-030-74513-4_3

3.1   Introduction

With the rapid development of the application of DGs, the power electronic inverters are being widely utilized to overcome the difficulties, such as controlling the voltage amplitude and frequency with a fast dynamic response and zero steady-state errors. For the single-phase inverters, the full-bridge pulse width modulation (PWM) inverter, whose major requirement of its control system is to control the voltage to achieve the steady state with zero steady-state error and a fast dynamic response, is widely used either in grid-connected or islanded mode voltage regulators in the distributed power systems [1]. Due to its simplicity in implementation and wide control bandwidth, deadbeat control is widely used in the PWM inverters. However, the parametric sensitivity, which exists in the system due to the high-order plant controlled by the deadbeat controller, may reduce the system stability margin [2, 3].

The repetitive control strategy, on the other hand, shows an excellent control performance for periodic signals, which makes it very effective for suppressing the harmonics and emulating various network dynamic disturbance scenarios. However, this scheme shows several limitations, such as the low-accuracy tracking performance, slow transient response, and poor rejection of the aperiodic disturbances [4]. The discrete-time sliding-mode is suggested as another control strategy to ensure the output voltage quality, which shows the advantage of robustness, fast transient response and simple digital implementation. Moreover, it is able to provide direct control without any modulation schemes. Despite the aforementioned advantages, this technique shows several drawbacks, including the chattering phenomenon resulted from the actuator limitations or time discretization, which can deteriorate the control accuracy and performance of the system [5].

When tracking a sinusoidal reference, the proportional-resonant (PR) control strategy does not require decoupling structures and it is able to ensure the system with zero steady-state error at a frequency even with the variations of the circuit parameters. However, the poor dynamic performance and the requirement of the very high switching frequency limit the application of this control technique [6]. To overcome these drawbacks, the synchronous reference frame proportional-integral (SRF-PI) controller, which is a well-developed technique in three-phase PWM converters, can be applied to single-phase inverters. By incorporating the SRF-PI controller into the single-phase PWM inverters, the zero steady-state error can be achieved through the conventional PI controller acting in the synchronous reference frame [7]. In this control structure, two orthogonal signals are generated by the orthogonal signal generation (OSG) techniques with respect to the fundamental frequency of the single-phase signals, and then transformed into the synchronous reference frame (SRF). A conventional PI regulator, which followed by the coordinate transformation, regulates the synchronous reference frame control signals to ensure a zero steady-state error, and then, the signals are transformed back to the stationary reference frame by an inverse Park's transformation [8, 9].

By employing a single-loop instantaneous voltage feedback control, zero steady-state error can be achieved in the inverter output voltage. However, in the industrial applications, a typical Inductor (L), Capacitor (C) filter, which is usually incorporated into the PWM converter to suppress the harmonic contents of the output voltage from the inverter, may introduce a resonance peak and reduce the stability margin of the PWM inverters. To overcome this shortcoming, in renewable energy system applications, dual-loop control strategies can be applied to the PWM inverters. For these strategies, the inner loops, which are usually the current loops, use the current of the filter inductor or capacitor current as the feedback signal to damp the resonant peak of the LC filter, and the outer loops, which are usually the voltage loops, use the filter capacitor voltage as the reference signal to regulate the output voltage [10, 11].

In [12], a dual-loop control strategy based on the hybrid reference frame is proposed, which adopts a capacitor voltage shaping loop with the SRF-PI controller and a capacitor current shaping loop in the stationary reference frame (here named HRF-based $v + i_c$ control strategy). This control strategy is able to achieve the steady-state with zero steady-state error and actively damp the resonance peak of the LC filter. However, the control parameter design method of the HRF-based $v + i_c$ control strategy is not complete in [12] since it neglects the impact of control delay, which is mainly brought by the computation and pulse width modulated (PWM) delays.

This chapter aims to provide a systematic parameter design guideline for the single-phase inverters using HRF-based $v + i_c$ control strategy with the effect of control delay. The dual-loop control strategy is presented in the hybrid reference frame for stand-alone single-phase inverters, which applies a capacitor voltage control loop in the synchronous reference frame and a capacitor current shaping loop in the stationary reference frame. The mathematical model of the single-phase inverter with this strategy is established under control delay scenario. A systematic method for designing SRF-PI controller and current controller is proposed for the HRF-based $v + i_c$ control strategy with consideration of the control delay. This method designs a satisfactory region, which is specified by the phase margin (PM) from 30° to 60° and the gain margin (GM) greater than 3 dB with the consideration of control delay of 150 μs. With this predefined region, the control parameters of the SRF-PI controller and the current controller can be easily obtained, and it is more convenient and explicit to optimize the system performance according to the satisfactory region.

This chapter is organized as follows. In Sect. 3.2, the control structure of the single-phase inverters with HRF-based $v + i_c$ control strategy and the equivalent model of the SRF-PI controller in the stationary reference frame are presented. The mathematic model of the HRF-based $v + i_c$ control strategy is established as well. In Sect. 3.3, a systematic parameter design guideline for the standalone single-phase inverters using HRF-based $v + i_c$ control strategy with consideration of control delay is proposed, which is conducted by specifying the satisfactory region of the stability indexes according to the stability margin. In Sects. 3.4 and 3.5, the simulation and experimental results are presented to validate the effectiveness of the proposed design approach. Finally, Sect. 3.6 concludes this chapter.

3.2   Control Structure of the Single-Phase Inverters

3.2.1   Control Structure of the Single-Phase Inverter with the HRF-Based Control Strategy

Figure 3.1 illustrates the control structure of the single-phase inverter using the HRF-based $v + i_c$ control strategy operated in the islanded mode. As shown in Fig. 3.1, for the power circuit of the single-phase inverter, an insulated-gate bipolar transistor (IGBT) full-bridge configuration, followed by an LC filter, is set as a VSI to produce PWM sinusoidal voltage V_{inv}. A linear load is in parallel with the capacitor. In the power circuit of the single-phase inverter, r_L denotes the series resistance of the inductor, i_l denotes the inductor current, i_c denotes the capacitor current, i_0 denotes the load current, and v_c denotes the capacitor voltage of the LC filter.

Meanwhile, the HRF-based $v + i_c$ control strategy for the full-bridge single-phase inverter is presented in Fig. 3.1 as well, which includes an SRF-PI voltage controller to regulate the output voltage and a capacitor current loop in the stationary reference frame to provide active damping and fast dynamic response. As shown in Fig. 3.1, it can be observed that the capacitor voltage and its orthogonal signal are transformed into synchronous frame by using Park transformation, which is followed by a PI controller. It should be mentioned that the orthogonal signal is generated by an all-pass filter, which causes a 90° phase delay at the fundamental frequency ω_f and

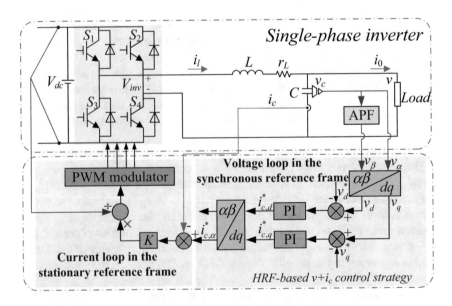

Fig. 3.1 The control structure of the single-phase inverter using the HRF-based $v + i_c$ control strategy [12, 13]

has unity-gain magnitude for all frequency ω with respect to the capacitor voltage. The structure of a first-order Active Power Filter (APF) is illustrated in Fig. 3.2 [12].

Since only α-axis quantities belong to the real system, the α-axis signal is fed forward to shape the voltage loop and at the same time, as the reference signal of the inner current loop when the HRF-based $v + i_c$ control strategy is conducted. It should be noted that the capacitor current (i_c) is selected as the feedback signal of the inner current loop. And the controller of the inner current loop is a proportional controller instead of a PI controller, which is popular in current feedback control. By applying a proportional controller, the phase delay problem can be easily solved compared to the PI controller and it is able to accelerate the dynamic response of the system.

3.2.2   Analysis of the SRF-PI Controller

Since the voltage loop works in the synchronous reference frame, which blocks the analysis of the whole closed-loop system and the appropriate design of the control parameters, it is essential to establish an equivalent model of the SRF-PI controller in the stationary reference frame. In [12], the stationary reference frame equivalent of the SRF-PI controller is derived by Monfared et al, which gives a better insight on the single-phase inverter using the HRF-based $v + i_c$ control strategy, and has a significant effect on parameter design and stability analysis of the system.

The block diagram of the SRF-PI controller is shown in Fig. 3.3, where the $G_{PI}(s)$ denotes the transfer function of the PI controller, that is, $G_{PI}(s) = K_p + K_i/s$. With two inputs and two outputs, which are both in the stationary reference frame, the equivalent of the structure shown in Fig. 3.3 can be written in time-domain as [12]:

$$
\begin{bmatrix} i_\alpha^*(t) \\ i_\beta^*(t) \end{bmatrix} = \begin{bmatrix} \cos(\omega_f t) & -\sin(\omega_f t) \\ \sin(\omega_f t) & \cos(\omega_f t) \end{bmatrix}
$$
$$
\left\{ \begin{bmatrix} G_{PI}(t) & 0 \\ 0 & G_{PI}(t) \end{bmatrix} * \left\{ \begin{bmatrix} \cos(\omega_f t) & \sin(\omega_f t) \\ -\sin(\omega_f t) & \cos(\omega_f t) \end{bmatrix} \begin{bmatrix} v_\alpha(t) \\ v_\beta(t) \end{bmatrix} \right\} \right\} \tag{3.1}
$$

where * denotes the convolution.

Taking the Laplace transform from both sides of (3.1) and substituting the transfer function of PI controller yield:

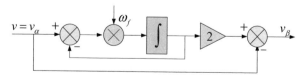

Fig. 3.2 The structure of a first-order APF [12]

Fig. 3.3 The block
diagram of the SRF-PI
controller [12]

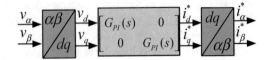

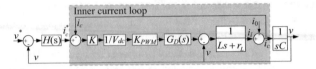

Fig. 3.4 Control scheme of the HRF-based $v + i_c$ control strategy in the stationary reference frame

$$i_\alpha^*(s) = \frac{a_3 s^3 + a_2 s^2 + a_1 s + a_0}{s^3 + \omega_f s^2 + \omega_f^2 s + \omega_f^3} v_\alpha(s) = H(s) v_\alpha(s) \tag{3.2}$$

where

$$\begin{cases} a_3 = K_p & a_2 = K_p \omega_f + K_i \\ a_1 = K_p \omega_f^2 + 2\omega_f K_i & a_0 = K_p \omega_f^3 - \omega_f^2 K_i \end{cases} \tag{3.3}$$

Hence, the $H(s)$ is the transfer function of the SRF-PI controller in the stationary reference frame, which has a significant influence on facilitating the analysis of the whole system and designing the control parameters of the inverter.

3.2.3   Mathematic Model of the HRF-Based Control Strategy

Figure 3.4 illustrates the control scheme of the HRF-based $v + i_c$ control strategy in the stationary reference frame. K_{pwm} is the transfer function of the PWM inverter, which is defined as V_{dc}/V_{tri}, where V_{dc} is the amplitude of the input dc voltage and V_{tri} is the amplitude of the triangular carrier signal. $G_D(s)$ denotes the transfer function of the control delay for this system, which can be expressed as $e^{-T_d \cdot s}$ with $T_d = 1.5/f_s$. Generally, the form of the $G_D(s)$ has three approximations, and to acquire a high bandwidth for the inverter in islanded mode, the approximation shown as (3.4) is preferred.

$$G_D(s) = e^{-T_d \cdot s} = \frac{1 - \frac{T_d}{2} s}{1 + \frac{T_d}{2} s} \tag{3.4}$$

Assuming that the linear load resistance is R and the i_0 in the block diagram of Fig. 3.4 can be expressed as

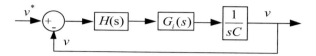

Fig. 3.5 The block diagram of the HRF-based $v + i_c$ control strategy

$$i_0 = v / R \tag{3.5}$$

In addition, according to [14], the amplitude of the triangular carrier signal V_{tri} is set as 1.0 pu. Hence, combining with (3.5) and $K_{pwm} = V_{dc}/V_{tri}$, the closed-loop transfer function of the inner current loop can be written as:

$$\frac{i_c}{i_c^*} = G_i(s) = \frac{K \cdot 1/V_{dc} \cdot K_{pwm} G_D RCs}{LRCs^2 + KG_D RCs + r_L RCs + Ls + r_L + R}$$

$$= \frac{KG_D RCs}{LRCs^2 + KG_D RCs + r_L RCs + Ls + r_L + R} \tag{3.6}$$

Hence, the block diagram of the HRF-based $v + i_c$ control strategy can be simplified as Fig. 3.5.

From Fig. 3.5, the open-loop transfer function of the block diagram of the HRF-based $v + i_c$ control strategy can be expressed as:

$$G_{open}(s) = H(s) \cdot \frac{K \cdot G_D \cdot R}{LRCs^2 + KG_D RCs + r_L RCs + Ls + r_L + R} \tag{3.7}$$

With the above analysis, it is evident that the design of the control parameters is of vital significance to ensure the stability and optimal performance of the standalone inverter system.

3.3   Step-by-Step Parameter Design

Since no parameter design guideline has been proposed for the single-phase inverters with HRF-based $v + i_c$ control strategy under control delay consideration, in this section, the design method of the control parameters is presented in detail. The presented approach is conducted by specifying the available region of the stability indexes, which is obtained according to the constraint of stability margin. With the particular region, the satisfactory stability indexes can be determined and the control parameters can be calculated with the selected stability indexes.

Figure 3.6 illustrates the Bode diagram of $H(s)$ for different values of K_i with $K_p = 1$. It can be observed that the value of K_i has no influence on the magnitude and phase frequency properties of the $H(s)$ at the high frequency range. Since the stability margins, corresponding to the phase margin (PM) and the gain margin (GM), are

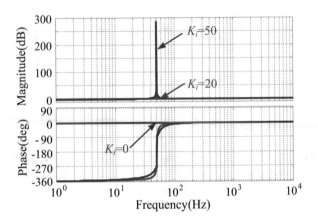

Fig. 3.6 The Bode diagram of the $H(s)$ with different values of K_i

both defined at the high frequencies, K_i can be selected as zero to simplify the $H(s)$, which yields:

$$H(s) \approx K_p \tag{3.8}$$

Hence, the open loop transfer function of the HRF-based $v + i_c$ control strategy can be simplified as:

$$G_{open}(s) = \frac{K_p \cdot K \cdot G_D \cdot R}{LRCs^2 + KG_D RCs + r_L RCs + Ls + r_L + R} \tag{3.9}$$

When designing a stable system, the phase should be above $-180°$ when the magnitude curve across 0 dB at the crossover frequency f_c on the Bode diagram to ensure PM > 0, and the magnitude should be below 0 dB when the phase curve across $-180°$ at the frequency f_g on the Bode diagram to ensure GM > 0. The expressions of the PM and GM can be written as [13, 14]:

$$PM = 180° + \arctan \angle G_{open}(j\omega)\big|_{\omega = 2\pi f_c} \tag{3.10}$$

$$GM = -20\lg \big|G_{open}(j\omega)\big|_{\omega = 2\pi f_g} \tag{3.11}$$

Substituting $s = j\omega$ into (3.9) yields:

$$G_{open}(j\omega) = \frac{K_p KR(1 - jA_1)}{A_2 + jA_3} \tag{3.12}$$

where the parameters A_1, A_2 and A_3 are denoted as:

$$A_1 = \frac{T_d}{2}\omega \tag{3.13}$$

$$A_2 = r_L + R - LRC\omega^2 + KRC\frac{T_d}{2}\omega^2 - r_L RC\frac{T_d}{2}\omega^2 - L\frac{T_d}{2}\omega^2 \tag{3.14}$$

$$A_3 = KRC\omega + r_L RC\omega + L\omega + r_L\frac{T_d}{2}\omega + R\frac{T_d}{2}\omega - LRC\frac{T_d}{2}\omega^3 \tag{3.15}$$

Hence, the phase angle and the magnitude of the open-loop transfer function can be expressed as:

$$\angle G_{open}(j\omega) = \arctan\frac{A_1 A_2 + A_3}{A_1 A_3 - A_2} \tag{3.16}$$

$$\left|G_{open}(j\omega)\right| = K_p KR\sqrt{\frac{A_1^2 + 1}{A_2^2 + A_3^2}} \tag{3.17}$$

Since the phase curve crosses $-180°$ at f_g, it can be obtained that

$$\left(A_1 A_2 + A_3\right)\Big|_{\omega=2\pi f_g} = 0 \tag{3.18}$$

Therefore, according to (3.18) and combining with (3.13), (3.14), and (3.15), the parameter K can be calculated as:

$$K = \frac{-L - T_d\left(r_L + R\right) - CRr_L + B_1 f_g^2}{CR + \pi^2 CRT_d^2 f_g^2} \tag{3.19}$$

where the variable B_1 is denoted as:

$$B_1 = \pi^2 r_L CRT_d^2 + \pi^2 T_d^2 L + 4\pi^2 CLRT_d \tag{3.20}$$

From (3.19), it can be inferred that K is proportional to f_g, which means that once the appropriate f_g is selected, the control parameter K can be easily determined.

Using the parameters listed in Table 3.1, substituting (3.16) and (3.19) into (3.10), the relationship among the PM, f_c, and f_g can be obtained, which is illustrated in Fig. 3.7. In industrial application, the PM is preferred to be $30° \sim 60°$ to achieve

Table 3.1 Parameters of the single-phase inverter

Symbol	Values
DC Link voltage (V_{dc})	50 V
Sampling and switching period(T_s)	100 μs
Fundamental frequency (ω_0)	100π rad/s
Filter inductance (L)	4000 μH
Filter capacitance (C)	2.2 μF
ESR of the inductor (r_L)	0.1 Ω
Control delay (T_d)	150 μs

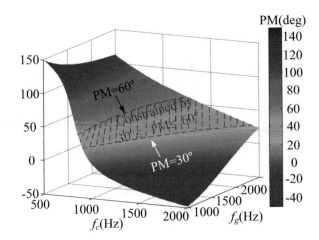

Fig. 3.7 The relationship of f_c, f_g, and PM

a good tradeoff between the system dynamic response and the requirement of a strong robustness. According to this, when considering the control delay, which is 150 μs in this paper, the available region constrained by PM = 30° and PM = 60° can be specified as the shadow shown in Fig. 3.7.

Similarly, at the crossover frequency f_c, the magnitude of the system can be written as

$$\left|G_{open}\left(s\right)\right|_{s=j2\pi f_c} = 1 \tag{3.21}$$

Substituting (3.21) into (3.17), it can be obtained that:

$$K_p = \frac{\sqrt{D_1^2 + D_2^2}}{KR\sqrt{\pi^2 T_d^2 \omega_c^2 + 1}} \tag{3.22}$$

where the parameters D_1 and D_2 are denoted as:

$$D_1 = \left[2\pi L + \left(r_L + R\right)\pi T_d + 2\pi\left(r_L + K\right)CR\right]f_c - 4\pi^3 CLRT_d f_c^3 \tag{3.23}$$

$$D_2 = r_L + R - 2\pi^2 T_d L f_c^2 - 4\pi^2 CLR f_c^2 + 2\pi^2\left(K - r_L\right)CRT_d f_c^2 \tag{3.24}$$

From (3.22), it can be concluded that K_p is related to K and f_c. Hence, once the appropriate K and f_c are selected, the control parameter K_p can be determined.

Substituting (3.17), (3.19), and (3.22) into (3.11), the relationship among the GM, f_c, and f_g can be obtained, which is illustrated in Fig. 3.8 to give a straightforward view. To ensure the stability of the system, the satisfactory region about GM is constrained by GM = 3 dB.

Fig. 3.8 The relationship
of f_c, f_g, and GM

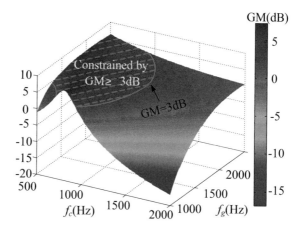

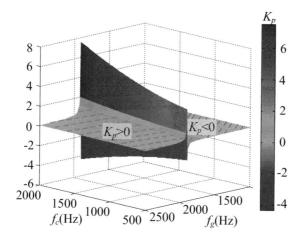

Fig. 3.9 The relationship of f_c, f_g and K_p

In Fig. 3.8, the satisfactory region constrained by GM ≥ 3 dB with control delay consideration is shown by the shadowed area, which is convenient to select the optimal f_c and f_g.

Moreover, K and K_p of the system should be positive. And the relationship of K_p, f_c, and f_g is illustrated in Fig. 3.9. From Fig. 3.9, it can be observed that when f_g is equal to 1910 Hz, different K_p can be obtained from 3.32 to 7.14 with the variations of f_c. When f_g > 1910 Hz, the values of K_p are between 0 and 2 and when f_g < 1910 Hz, the values of K_p are negative. For the control parameter K, since it is proportional to f_g, when f_g is greater than 1907, the value of K is greater than 0.

According to the above analysis, the satisfactory region of f_g and f_c can be determined, which is plotted in the shadow in Fig. 3.10. In Fig. 3.10, point A is selected

Fig. 3.10 The satisfactory region of f_g and f_c

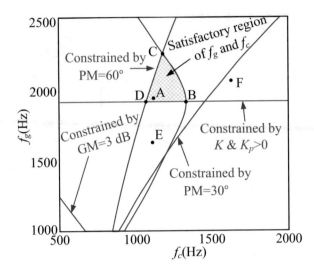

Table 3.2 System performance indexes

	Point A	Point B	Point C	Point D	Point E	Point F
f_c(Hz)	1110	1310	1170	1070	1170	1650
f_g(Hz)	1916	1910	2260	1910	1670	2120
K	0.89	0.34	30	0.33	-23	19
K_p	1.71	5.06	0.07	4.40	-0.06	0.12
PM(deg)	57.50	40.71	60.82	60.85	41.88	26.60
GM(dB)	4.04	3.04	3.00	4.25	3.94	1.54

in the satisfactory region with f_c = 1110 Hz and f_g = 1916 Hz. Substituting the selected f_c and f_g into (3.19) and (3.22), respectively, yields:

$$K = 0.89, K_p = 1.71 \tag{3.25}$$

Moreover, as shown in Fig. 3.10, points B, C, and D are selected on the edge of the satisfactory region. Points E and F are selected out of the region, which are out of the constraint line about K and K_p and the constraint line about PM and GM, respectively.

The corresponding system performance indexes of points B, C, D, E, and F are shown in Table 3.2. From the indexes of points B, C, and D, it can be obtained that the max range of f_c is from 1070 to 1310 Hz, and with the increase of f_g, the range of f_c decreases. Point E satisfies the specifications of the stable margin with PM = 41.88° and GM = 3.94 dB, but the values of K and K_p are negative. Point F results positive K and K_p, but the PM and GM are smaller than 30° and 3 dB, respectively, which are consistent with the theoretical analysis.

For the control parameter K_i, as mentioned earlier, it mainly affects the magnitude at the fundamental frequency instead of the stability of the system. Hence, K_i is selected as 10 to ensure the zero steady-state error at the fundamental frequency.

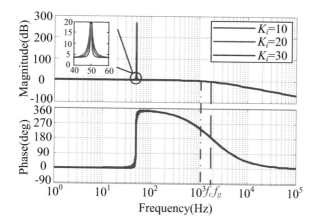

Fig. 3.11 The Bode diagram of the open loop transfer function of the system with $K = 0.89$, $K_p = 1.71$, and different K_i

Figure 3.11 illustrates the Bode diagram of the open loop transfer function of the system with $K = 0.89$, $K_p = 1.71$, and different K_i. It can be observed that all the cases are able to guarantee the stability of the system with zero steady-state error and $K_i = 10$ has the optimal filtering capacity around the fundamental frequency. Moreover, when $K_i = 10$, the crossover frequency f_c of the system is 1110 Hz, f_g is 1916 Hz, PM is 57.50°, and GM is 4.04 dB, which all satisfy the aforementioned specifications.

According to the above analysis, the step-by-step parameter design method can be summarized as:

Step 1: Specify the phase margin and the gain margin of the system, determine the satisfactory region of f_c and f_g according to (3.10), (3.11), and define $K > 0$ and $K_p > 0$.

Step 2: Choose the optimal values of f_c and f_g from the satisfactory region.

Step 3: Calculate the control parameter K and K_p with the selected f_c and f_g according to (3.19) and (3.22).

Step 4: Choose a proper K_i and validate the selected K_i.

Figure 3.12 illustrates the variation tendencies of the PM and GM of the systems using the HRF-based $v + i_c$ control strategy and single-loop voltage control strategy, respectively, with respect to the system delay T_d variation. It can be observed that the PM and GM of both systems decrease when T_d increases. However, when T_d increases to 180 μs, the PM and GM of the system using single-loop voltage control strategy are smaller than 30° and 3 dB, respectively, while the system using HRF-based $v + i_c$ control strategy remains a good stability margin with a PM of about 44° and a GM of about 4 dB.

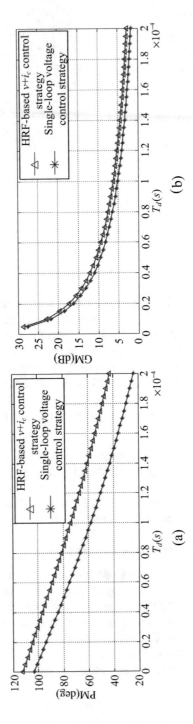

Fig. 3.12 The variation tendencies of the PM and GM of the systems

3.4 Stability Analysis Using EMTP-ATP Simulation

Figure 3.13 shows the dynamic performance of the voltage loop tracking controller with $T_d = 1.5T_s$ under different load conditions, where the effect of light load $R_L = 10$ Ohm and heavy load $R_L = 5$ Ohm are considered. It can be observed that under light load conditions, a slight oscillation amplification is generated when the single-loop control scheme is applied. This oscillation is significantly damped when the dual-loop control scheme is utilized, as shown in Fig. 3.13b. In case of heavy load conditions, the closed-loop stability is ensured under both single-loop and dual-loop control schemes.

In case of heavy load condition, the inner loop controller gain can be increased to improve the tracking performance, while maintaining the closed-loop stability of the inverter system, since the load impedance provides auxiliary damping. Next, the load resistance is assumed to be 5 Ohm, and the effect of current loop is further studied.

As shown in Fig. 3.14, with an increase of the inner loop gain, the tracking error is remarkably reduced, compared to the simulation results in Fig. 3.13, either in the single-loop or dual-loop control schemes.

Next, in order to check the robustness of the control scheme, the parameter uncertainties are tested in terms of closed-loop stability characteristics, where the variations of the filter inductance, capacitance, and load resistance are considered.

Figure 3.15 shows the sensitivity analysis under the variation of filter inductance; the system instability may occur when $\triangle L = -15\%$, when the single-loop control scheme is adopted. This tendency would be improved in case of the dual-loop control scheme. However, instability would also occur with a slower divergence rate. The reduction of the filter inductance may imply an excessive high controller gain, thus causing system instability due to insufficient damping. When $\triangle L = -10\%$, the divergence rate is much slower compared to the case of $\triangle L = -15\%$. Whereas, the system instability would also occur when the single-loop control scheme is applied. Nevertheless, when the dual-loop control scheme is applied, system stability is ensured under this scenario. When $\triangle L = -5\%$, the damping is remarkably enhanced, thus the stability of the closed-loop system can be guaranteed, with a little transient oscillation in case of the single-loop control scheme. When $\triangle L = +20\%$, the closed-loop stability is ensured under both control strategies due to enhanced damping effect. It can be concluded from Fig. 3.15 that, the reduction of filter inductance may generate transient oscillation or instability due to insufficient system damping, whereas an increase of filter inductance would enhance system stability.

Figure 3.16 shows the sensitivity analysis under the variation of filter capacitance; the system instability may occur when $\triangle C = +20\%$, when the single-loop control scheme is adopted. This tendency would be improved in case of the dual-loop control scheme. When $\triangle C = +10\%$, the damping is remarkably enhanced, thus the stability of the closed-loop system can be guaranteed, with a little transient oscillation in case of the single-loop control scheme. When $\triangle C = -20\%$, the

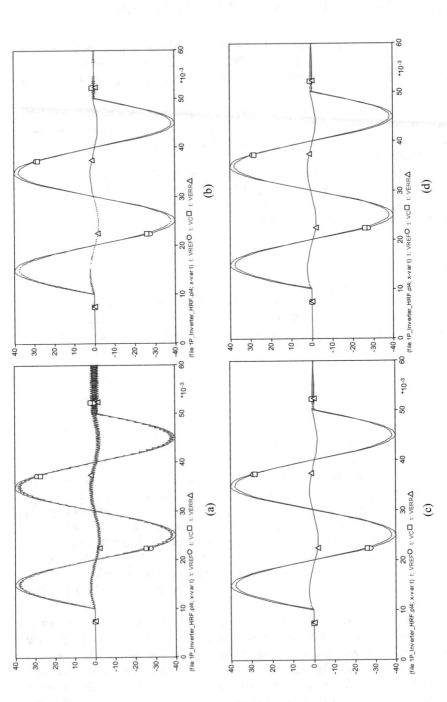

Fig. 3.13 Dynamic performance of the voltage loop tracking controller with $T_d = 1.5T_s$ under different load conditions. (**a**) Single-loop control with $R_L = 10$ Ohm; (**b**) Dual-loop control with $R_L = 10$ Ohm; (**c**) Single-loop control with $R_L = 5$ Ohm; (**d**) Dual-loop control with $R_L = 5$ Ohm

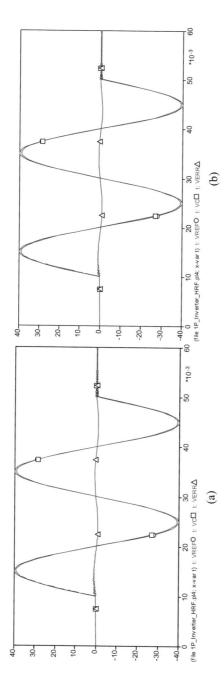

Fig. 3.14 Dynamic performance of the voltage loop tracking controller with $T_d = 1.5T_s$ and increased inner loop controller gain. (**a**) Single-loop control with $R_L = 5$ Ohm; (**b**) Dual-loop control with $R_L = 5$ Ohm

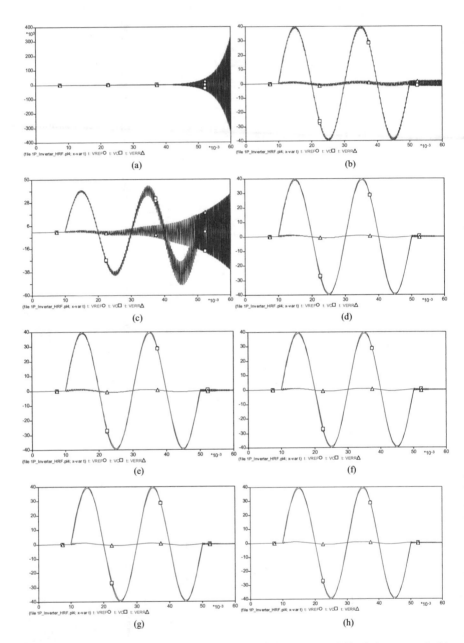

Fig. 3.15 Sensitivity analysis under the variation of filter inductance. (**a**) Single-loop control with $\triangle L = -15\%$; (**b**) Dual-loop control with $\triangle L = -15\%$; (**c**) Single-loop control with $\triangle L = -10\%$; (**d**) Dual-loop control with $\triangle L = -10\%$; (**e**) Single-loop control with $\triangle L = -5\%$; (**f**) Dual-loop control with $\triangle L = -5\%$; (**g**) Single-loop control with $\triangle L = +20\%$; (**h**) Dual-loop control with $\triangle L = +20\%$

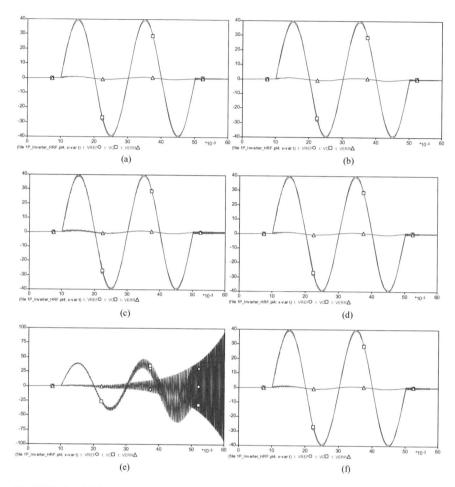

Fig. 3.16 Sensitivity analysis under the variation of filter capacitance. (**a**) Single-loop control with $\triangle C = -20\%$; (**b**) Dual-loop control with $\triangle C = -20\%$; (**a**) Single-loop control with $\triangle C = +10\%$; (**b**) Dual-loop control with $\triangle C = +10\%$; (**c**) Single-loop control with $\triangle C = +20\%$; (**d**) Dual-loop control with $\triangle C = +20\%$

closed-loop stability is ensured under both control strategies due to enhanced damping effect. It can be concluded from Fig. 3.16 that the increase of filter capacitance may generate transient oscillation or instability due to insufficient system damping, whereas a decrease of filter capacitance would enhance system stability.

Figure 3.17 shows the sensitivity analysis under the variation of load resistance; the system instability may occur when $\triangle R_L = +10\%$, when the single-loop control scheme is adopted. However, the stability margin would be enhanced with a little transient oscillation phenomenon with the dual-loop control scheme. In case of $\triangle R_L = -10\%$, the stability margin is improved under both single-loop and dual-loop control schemes. This effect is reasonable since a reduction of load side

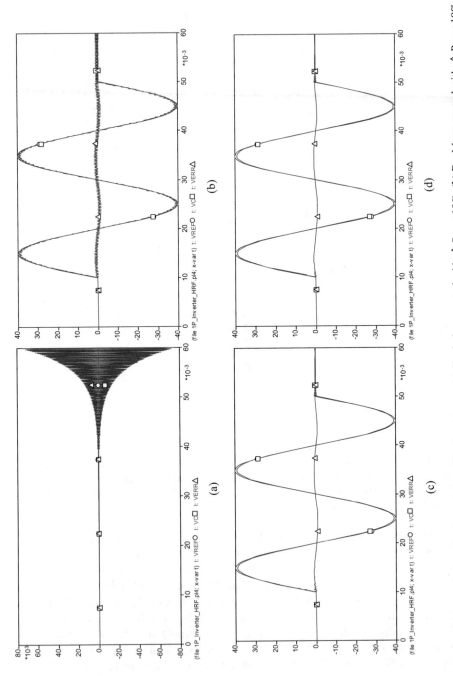

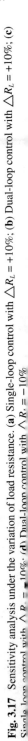

Fig. 3.17 Sensitivity analysis under the variation of load resistance. (**a**) Single-loop control with $\triangle R_L = +10\%$; (**b**) Dual-loop control with $\triangle R_L = +10\%$; (**c**) Single-loop control with $\triangle R_L = -10\%$; (**d**) Dual-loop control with $\triangle R_L = -10\%$

resistance implies a higher damping, thus; improved stability characteristics can be ensured.

3.5 Experimental Evaluation

To evaluate the effectiveness of the HRF-based $v + i_c$ control strategy with the designed control parameters, a downscaled single-phase inverter system is set up, which consists of a power dc source of 50 V, a full-bridge Metal Oxide Silicon Field Effect Transistor (MOSFET) power module, an LC filter, gate drivers and sensors. The control algorithm is implemented in TMS320F28335 controller and the reference signal of d-axis in synchronous reference frame is set to 40 V.

In the first case, the steady-state performance and transient performance of the inverter with $K = 0.89$, $K_p = 1.71$, and $K_i = 10$ are investigated, which are shown in Figs. 3.18 and 3.19, respectively. From Fig. 3.18, it can be observed that the system achieves an excellent steady-state performance and the reference voltage can be tracked accurately by the output voltage of the system. The harmonic spectrum is shown as well. From the fast Fourier transformation (FFT) analysis, it can be obtained that the total harmonic distortion (THD) under this situation is 3.68%, which shows a satisfied performance under this scenario.

Figure 3.19 illustrates the transient performance of the inverter with $K = 0.89$, $K_p = 1.71$, and $K_i = 10$. Figure 3.19a shows the transient waveforms in response to turning on the inverter. As shown in Fig. 3.19a, the system tracks the reference voltage in 4 ms, with slight overshoot of about 2 V during the transient process, which shows a fast dynamic response and a strong robustness. Figure 3.19b shows the transient response of the inverter from nominal load to 50% nominal load condition. It can be observed that during the transient process, slight ripples occur and the

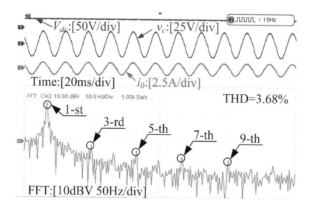

Fig. 3.18 Experimental waveforms of the steady-state performance of the inverters when $K = 0.89$, $K_p = 1.71$, and $K_i = 10$ under nominal load ($R = 20\ \Omega$)

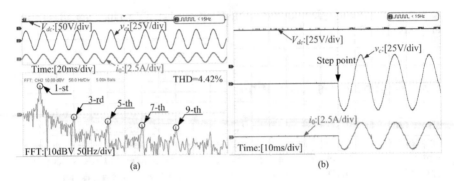

Fig. 3.19 Experimental waveforms of transient performance of the inverter when $K = 0.89$, $K_p = 1.71$, and $K_i = 10$ (**a**) transient waveforms in response to turning on the inverter; (**b**) transient waveforms undergoes nominal load to 50% nominal load step change

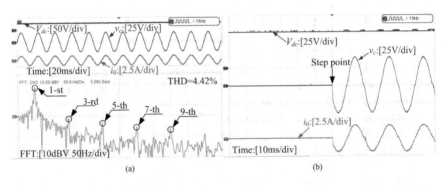

Fig. 3.20 Experimental waveforms of the inverter when $K = 19$, $K_p = 0.12$, and $K_i = 10$ (**a**) the steady-state performance; (**b**) transient performance in response to turning on the inverter

system achieves the steady-state in 3 ms, which shows a fast transient performance as well.

Then, the steady-state performance and transient response of the inverter with $K = 19$, $K_p = 0.12$, and $K_i = 10$, which corresponding to the point F in Sect. 3.3, are investigated in Fig. 3.20 (a) and (b), respectively. From Fig. 3.20, it can be observed that the THD of the system under the steady state is 4.42% and the response time is about 7 ms, which means that compared to the case when $K = 0.89$, $K_p = 1.71$, and $K_i = 10$, the parameters corresponding to point F can also achieve the steady-state, but with a relatively poor performance on the stability and the transient response.

Figure 3.21 illustrates the experimental waveforms of the steady-state performance of the inverter when K_p increases. In Fig. 3.21a, K_p is increased to 2, and it can be observed that slight oscillations appear in the capacitor voltage and the load current with the voltage THD of about 4.00% under this scenario. Figure 3.21b shows that, when K decreases, the capacitor voltage fails to track the reference voltage accurately and the corresponding THD is 4.15%. When K increases, which is

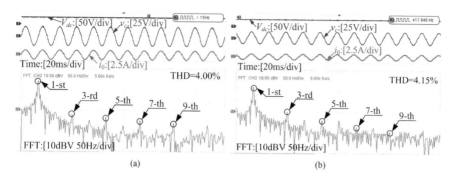

Fig. 3.21 Experimental waveforms of the steady-state performance of the inverter when (**a**) $K = 0.89$, $K_p = 2$, and $K_i = 10$; (**b**) $K = 0.7$, $K_p = 1.71$, and $K_i = 10$

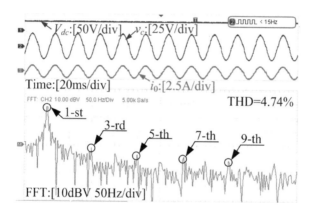

Fig. 3.22 Experimental waveforms of the steady-state performance of the inverter with $K = 1.2$, $K_p = 1.71$, and $K_i = 10$

shown in Fig. 3.22, the waveforms of v_c and i_0 are distorted with the voltage THD of 4.74%, which means that the inverter is slightly oscillating.

Finally, the steady-state performance of the inverters with $K = 0.89$, $K_p = 1.71$, and $K_i = 10$ under RL load and nonlinear load is studied. Figure 3.23a depicts that the system is able to achieve the steady state under RL load conditions and the THD of the system under this situation is 4.09%. Figure 3.23b shows the system performance under nonlinear load condition. It can be observed that the system achieves the steady state with the THD of 5.15%, which shows a good performance with the designed parameters.

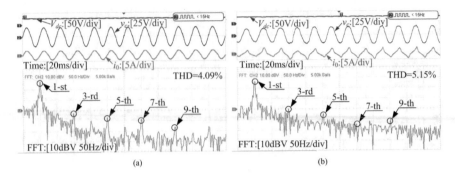

Fig. 3.23 Experimental waveforms of the steady-state performance of the inverters with $K = 0.89$, $K_p = 1.71$, and $K_i = 10$ under (**a**) RL load ($R = 10\,\Omega$, $L = 3.8$ mH); (**b**) Nonlinear load ($L = 3.8$ mH, $C = 2000\,\mu$F, $R = 50\,\Omega$)

3.6   Conclusion

This chapter presents a classical controller synthesis and parameter selection criterion for the single-phase inverter controlled in standalone mode, which applies a
capacitor voltage loop with the PI controller in the SRF and a capacitor current loop
with the proportional controller in the stationary reference frame. Taking into
account the effect of control delay, the control structure of the HRF-based $v + i_c$
control strategy is analyzed, and the parameter design guideline for the SRF-PI
controller for the voltage loop and proportional controller for the current loop is
presented. Moreover, the mathematical model of the HRF-based $v + i_c$ control strategy is established considering the effect of control delay. By specifying the phase
margin and the gain margin of the system, the satisfactory region of the stability
indexes can be obtained. And the control parameters can be calculated with the
stability indexes selected from the region. With the designed parameters, the system
has a fast transient response and a strong robustness against the time delay. The
simulation results obtained from the EMTP software are presented to test the system
robustness under power-stage parameter uncertainties. Moreover, the experimental
results are presented to validate the effectiveness of the parameter design method,
which can be widely applied for the single-phase PWM inverters of DGs in the
islanded operating conditions.

References

1. Guo, X., & Guerrero, J. M. (2016). General unified integral controller with zero steady-state
   error for single-phase grid connected inverters. *IEEE Transactions on Smart Grid, 7*(1), 74–83.
2. Song, W. S., Ma, J. P., Zhou, L., & Feng, X. Y. (2016). Deadbeat predictive power control of
   single-phase three-level neutral-point-clamped converters using space-vector modulation for
   electric railway traction. *IEEE Transactions on Power Electronics, 31*(01), 721–732.

3. Pichan, M., Rastegar, H., & Monfared, M. (2017). Deadbeat control of the stand-alone four-leg inverter considering the effect of the neutral line inductor. *IEEE Transactions on Industrial Electronics, 64*(4), 2592–2601.
4. Yang, S., Wang, P., Tang, Y., & Zhang, L. (2017). Explicit phase lead filter design in repetitive control for voltage harmonic mitigation of VSI-based islanded microgrids. *IEEE Transactions on Industrial Electronics, 64*(1), 817–826.
5. Dehkordi, N. M., Sadati, N., & Hamzeh, M. (2017). A robust backstepping high-order sliding mode control strategy for grid-connected DG units with harmonic/interharmonic current compensation capability. *IEEE Transactions on Sustainable Energy, 8*(2), 561–572.
6. Kuperman, A. (2015). Proportional-resonant current controllers design based on desired transient performance. *IEEE Transactions on Power Electronics, 30*(10), 5341–5345.
7. Roshan, A., Burgos, R., Baisden, A. C., Wang, F., & Boroyevich, D. (2007). A D-Q frame controller for a full-bridge single phase inverter used in small distributed power generation systems. In *Proc. IEEE APEC* (pp. 641–647).
8. Bahrani, B., Rufer, A., Kenzelmann, S., & Lopes, L. (2011). Vector control of single-phase voltage source converters based on fictive axis emulation. *IEEE Transactions on Industry Applications, 47*(2), 831–840.
9. Chen, J., Zhang, W., Chen, B., & Ma, Y. (2016). Improved vector control of brushless doubly fed induction generator under unbalanced grid conditions for offshore wind power generation. *IEEE Transactions on Energy Conversion, 31*(1), 293–302.
10. Vidal, A., Freijedo, F. D., Yepes, A. G., Malvar, J., Lopez, O., & Doval-Gandoy, J. (2014). Transient response evaluation of stationary-frame resonant current controllers for grid-connected applications. *IET Power Electronics, 7*(7), 1714–1724.
11. Xu, S. G., Wang, J. P., & Xu, J. P. (2013). A current decoupling parallel control strategy of single-phase inverter with voltage and current dual closed-loop feedback. *IEEE Transactions on Industrial Electronics, 60*(4), 1306–1313.
12. Monfared, M., Golestan, S., & Guerrero, J. M. (2014). Analysis, design, and experimental verification of a synchronous reference frame voltage control for single-phase inverters. *IEEE Transactions on Industrial Electronics, 61*(1), 258–269.
13. Han, Y., Jiang, A., Coelho, E. A. A., & Guerrero, J. M. (2018). Optimal performance design guideline of hybrid reference frame based dual-loop control strategy for standalone single-phase inverters. *IEEE Transactions on Energy Conversion, 33*(2), 730–740.
14. Han, Y., Li, Z., Yang, P., Wang, C., Xu, L., & Guerrero, J. (2017). Analysis and design of improved weighted average current control strategy for LCL-type grid-connected inverters. *IEEE Transactions on Energy Conversion, 32*(3), 941–952.

Chapter 4
Nonlinear Stability Analysis of Digital Controlled Single-Phase Standalone Inverter

In the last chapter, the modeling and controller synthesis approach of single-phase standalone inverter using classical control theory was discussed. However, for the single-phase converters with synchronous reference frame (SRF) control loops, little work has been done on the evaluation of the nonlinear approaches for stability analysis. In this chapter, the stability characteristics of a digital-controlled single-phase voltage source inverter (VSI) with SRF voltage control loop are investigated from the perspective of nonlinear system. The stability analysis is implemented using the discrete-time model defined by the stroboscopic map, which is derived using the state-space averaging (SSA) technique. Furthermore, two different nonlinear analysis methods, the Jacobian matrix method and Lyapunov exponent method, are adopted to analyze the fast-scale stability and the slow-scale stability of the PWM inverter under variations of control parameters; hence, the stability regions can be easily obtained analytically. The theoretical results indicate that, for the established stroboscopic models, the Jacobian matrix method and the Lyapunov exponent method are mathematically equivalent, which means that the fast-scale stability and slow-scale stability of the studied single-phase standalone inverters are consistent, especially under linear load conditions. Simulation results obtained from EMTP software are presented to study the effect of control delay, load parameters, and controller gain on stability characteristics of the closed-loop system. In order to validate the theoretical analysis, the experimental results under resistive load, inductive-resistive load, and diode rectifier load conditions are presented, which also proves that the discrete-time model plus Jacobian matrix method or Lyapunov exponent method is capable to accurately investigate the stability boundaries of a voltage source converter operating in the standalone mode with SRF control loops.

© Springer Nature Switzerland AG 2022
Y. Han, *Modeling and Control of Power Electronic Converters for Microgrid Applications*, https://doi.org/10.1007/978-3-030-74513-4_4

4.1   Introduction

As the interface for single-phase grid or local loads, the single-phase voltage source inverters (VSIs) are widely used in various industrial fields, and play an important role in renewable energy systems including distributed generations (DGs) and microgrids (MGs), due to the increasing penetration of renewable energy in recent years. Single-phase VSIs can work either in grid-connected mode or stand-alone mode, and are closely combined with the pulse-width modulation (PWM) technique and digital control technologies. In general, the most common application of single-phase VSIs in stand-alone mode lies in off-grid power generation systems and power equipment like uninterrupted power supply (UPS). This kind of converter is normally designed with an Inductor (L), Capacitor (C) smoothing filter and closed-loop control structure to produce a stable sinusoidal output voltage of constant magnitude and frequency with fast dynamic response and zero steady-state error [1–4].

Apart from the conventional single or dual closed-loop control strategy based on the proportional-integral-derivative (PID) regulators, the deadbeat control, repetitive control, sliding mode control, and proportional-resonant (PR) control are the most frequently used control methods. The deadbeat control possesses excellent dynamic performance and wide control bandwidth due to the direct regulation of the inverter output voltage, but it is highly sensitive to system parameters and cannot remove the steady-state errors of the system. The repetitive control is mainly designed for systems with periodic output, and it is effective in suppressing the harmonics of the output voltage. However, poor rejection of aperiodic disturbance, slow dynamics, and low tracking accuracy normally limits the application of this control technique. The sliding mode control exhibits superiority in the dynamic behaviors, implementation simplicity, and less additional regulation. Despite these advantages, sliding mode control also suffers from the obvious flaw of dynamic tracking accuracy. The proportional-resonant (PR) control is well known for its capacity of effectively eliminating the steady-state error in tracking ac signals and applicability of instantaneous voltage control for single-phase VSIs, and the PR control scheme containing multiple resonant units is a prevalent method for harmonic compensation. But PR control is also constrained by the disadvantages of poor dynamic response to input changes and great sensitivity to deviations of sampling signals. In addition to these methods, some intelligent control approaches, such as adaptive control, neural network control, and fuzzy control, have also been presented in literatures. Generally, intelligent control methods are applied in practical applications for their advantages of strong robustness, low dependence on system parameters, and adaptive characteristics, which means that these approaches are suitable for nonlinear, time-varying, or delay systems. However, due to the high complexity, limited control precision, and the lack of complete analysis and design guidelines, further in-depth research is required to apply the intelligent control methods to practical applications.

Due to its advantage of realizing a zero steady-state error, the synchronous reference frame (SRF) control scheme for the single-phase VSI has attracted increasing interests by employing the conventional Proportional-Integral (PI) regulators in the SRF. To utilize this control technique, a fictitious second phase voltage is generated by the orthogonal-signal-generation (OSG) techniques to emulate a two-phase system, and the electrical signals are transformed to the SRF for effective control. In the last chapter, the modeling and controller synthesis approach of the single-phase standalone inverter using classical control theory was discussed. However, the switching converters are nonlinear systems in nature, whose nonlinear characteristics are originated from both the power circuits and the control systems. Despite the broad applicability for switching converters, linear analysis methods like transfer functions suffer from the drawbacks of poor description for the nonlinearity and the fast-scale dynamics of switching converters, and incomplete stability prediction. On the contrary, the nonlinear approaches are capable of illustrating the slow- and fast-scale stability of the switching converters directly and are well adapted to present the nonlinear phenomena such as bifurcation and chaos in switching converters.

Furthermore, the nonlinear approaches are suitable for analyzing the digital controlled switching converters reliably and accurately. In retrospect, the nonlinear control theory is first adopted to study the nonlinear phenomena in DC-DC converters and then extended to the other switching converters like PWM inverters and power factor correction (PFC) circuits. Several nonlinear control strategies for switching converters have been developed, such as the Lyapunov function-based and passivity-based control methods. These control strategies reserve the nonlinearities of the switching converters, but also ensure high control performances, including global stability, improved waveforms, zero steady-state error, and fast dynamics under linear or nonlinear load conditions. For stability analysis of switching converters, bifurcation diagram method, Lyapunov exponent method, and Jacobian matrix method are three main nonlinear approaches. These three methods are normally based on the discrete-time model, and for a certain system, they can be applied simultaneously. Bifurcation diagram method describes system stability through the bifurcation diagrams. To utilize this method, iterative calculation determined by the discrete-time model should be implemented to compute numerous values of state variables under different bifurcation parameters. By plotting the calculated state variable values with the corresponding bifurcation parameter values into graphs with certain rules, bifurcation diagram can be produced. In an ordinary bifurcation diagram, bifurcation parameter is on one axis as an independent variable and state value on the other axis as a dependent variable. The system is stable on the bifurcation parameter intervals where one certain bifurcation parameter value corresponds to only one state variable value. The system goes into the unstable period-n state on the bifurcation parameter intervals where one certain value of bifurcation parameter corresponds to n ($n \geq 2$) state variable values. And on the bifurcation parameter intervals where one certain bifurcation parameter value corresponds to infinite state variable values, the system operates in a chaotic state which is highly unstable [5–8].

Bifurcation diagram presents the stable and unstable parameter intervals, and the processes of bifurcations in a straightforward manner. However, to use this method, state variables are required to be precisely calculated in iterations, which is not available for all discrete-time models, and may lead to a huge computational burden. The Lyapunov exponent method depicts the system stability by employing the Lyapunov exponent. An n-dimensional system possesses n Lyapunov exponents, and system stability can be described by the maximum one. The stability criteria of the Lyapunov exponent method is, negative maximum Lyapunov exponents indicate that the system is stable, zero maximum Lyapunov exponents indicate that system operates in critical steady state, and positive maximum Lyapunov exponents indicate that the system is chaotic. The maximum Lyapunov exponent can be calculated by several approaches. By plotting the maximum Lyapunov exponent with the selected system parameter into graphs, the maximum Lyapunov exponent spectrums can be obtained. In maximum Lyapunov exponent spectrum, the stable and unstable parameter intervals can be demonstrated clearly. The principle of Jacobian matrix method is to determine system stability on the basis of the eigenvalues of Jacobian matrix at the fixed point of discrete-time model [9, 10].

The stability criteria of Jacobian matrix method can be expressed as: when all eigenvalues of Jacobian matrix at the fixed point of discrete-time model are located in the unit circle on the complex plane, system is stable, and when any eigenvalue lies outside the unit circle, system becomes chaotic. For the critical situations that some eigenvalues lie on the unit circle, but no eigenvalue lies outside it, system moves into the critical steady state. The Jacobian matrix method is carried out in one single switching cycle, which is a sufficiently short time period that can be defined as the so-called "fast scale." It reveals the system dynamic behavior which possesses low amplitude or high frequency close to the switching frequency, so the stability described by this feature is usually called the fast-scale stability of switching converters. On the contrary, the bifurcation diagram method and Lyapunov exponent method are both implemented in multiple successive switching cycles, which constitute a much longer time period that can be defined as the "slow scale," and they normally demonstrate the system dynamic property with a frequency that is much lower than the switching frequency, and the stability characteristics can be called the slow-scale stability of switching converters [11, 12].

The discrete-time model is, undoubtedly, required for stability analysis of switching converters using the bifurcation diagram method, the Lyapunov exponent method, or the Jacobian matrix method. Discrete-time models of switching converters are usually derived by using discrete maps. Depending on the mapping points, discrete maps mainly include stroboscopic map, synchronous switching map, asynchronous switching, and two-by-two map, while the stroboscopic map is the most popular one. In stroboscopic map, the state variables at the end point are derived by solving the state equations with state variables at the starting point within one switching cycle, which is equivalent to sampling the state variables with switching frequency. For stability analysis, stroboscopic model defined by the stroboscopic map, is proved to be reliable and accurate. However, the inherent piecewise-linear property of switching converters can bring great difficulties in calculating the exact solutions of the state equations during a switching cycle, especially for state equations of high order, which creates a limitation for stroboscopic map. In addition, a comparison of stroboscopic models derived by using the state-space average technique, and precisely solving the state equations is presented to confirm to study the nonlinear characteristics of switching converters with high switching frequency [13–16].

In this chapter, the detailed stability analysis of a digital controlled single-phase VSI with SRF voltage control is presented by employing nonlinear approaches. The stroboscopic model of the inverter is established by using the state-space averaging technique, and analyzed by Jacobian matrix method and Lyapunov exponent method under control parameters variations. The stability regions of the PWM inverter are obtained, and the analytical results show that, for the studied stroboscopic model, the fast-scale stability described by Jacobian matrix method is equivalent to the slow-scale stability determined by Lyapunov exponent method. Then, the simulation results obtained from EMTP software are presented to study the effect of control delay, load parameters, and controller gain on stability characteristics of the closed-loop system. Moreover, the experimental results under resistive load, inductive-resistive load, and diode rectifier load conditions are presented to validate the theoretical analyses. The presented approach can also be extended to other converter topologies either in the standalone or grid-connected operating mode.

4.2 Mathematical Modeling with SRF Voltage Control

The digital controlled single-phase VSI with SRF voltage control is shown in Fig. 4.1, which works in the stand-alone mode with an LC filter and the load Z. As shown in Fig. 4.1, the filter capacitor voltage plus the filter capacitor current

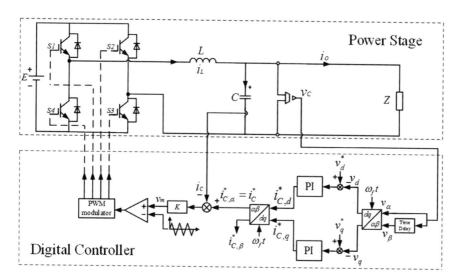

Fig. 4.1 System structure of the digital controlled single-phase VSI with SRF voltage control in stand-alone mode

feedback control strategy is applied in this inverter system. Specifically, to emulate a two-phase system, the filter capacitor voltage v_C is taken as the α-axis input v_α for the Park's transformation and a fictitious electrical signal generated by the time delay block serves as the β-axis input v_β.

In the stationary frame, the reference of v_C is defined as $v_C^* = V_{Cm}\cos(\omega_f t)$ with a fundamental cycle of $T_f = 2\pi/\omega_f$, so the time delay block delays v_C for one quarter of T_f in time domain to ensure that its output v_β is orthogonal to v_C. The d-axis reference voltage v_d^* in the SRF is then set to the desired magnitude V_{Cm}, and the q-axis reference voltage v_q^* is equal to zero. After the same PI control for the deviation of both v_d and v_q, the electrical signals are transformed back to stationary frame by inverse Park's transformation, and the α-axis output is subsequently taken as the reference for the filter capacitor current i_C in current loop, since only α-axis quantities correspond to the real system. In the current loop, the deviation of i_C is regulated by a proportional controller, and then modulation signal v_m is finally produced.

Main parameters of the studied PWM inverter with a reduced-scale power rating are: dc-link voltage $E = 50$ V, filter inductance and capacitance are 2 mH and 2.2 μF, respectively. The sampling and switching frequencies are both 20 kHz, and d-axis reference value is set as 40 V. Three important control parameters, proportional gain k_i and integral gain k_p of the PI controllers in voltage loop, and the proportional gain K in current loop, since their effect on the system stability is investigated in the following subsections.

4.2.1 System Modeling Under Resistive Load Condition

In this case, load Z in Fig. 4.1 is regarded as a linear resistor of 20 Ω, which is denoted by R. Since the PWM inverter in stand-alone mode is composed of power stage and digital controller, these two parts are modeled simultaneously in the following parts. The stroboscopic map used for the modeling is shown in Fig. 4.2.

For the power stage, the filter inductor current i_L and filter capacitor voltage v_C are considered as state variables. In one switching cycle, the state in which S1 and S3 are on, and S2 and S4 are off is defined as State 1, while the state in which S2 and S4 are on, and S1 and S3 are off is defined as State 2. State equation for State 1 is derived as

$$\dot{\mathbf{x}} = \mathbf{A}_1\mathbf{x} + E\mathbf{B}_1 \tag{4.1}$$

and state equation for State 2 is written as

$$\dot{\mathbf{x}} = \mathbf{A}_2\mathbf{x} + E\mathbf{B}_2 \tag{4.2}$$

where

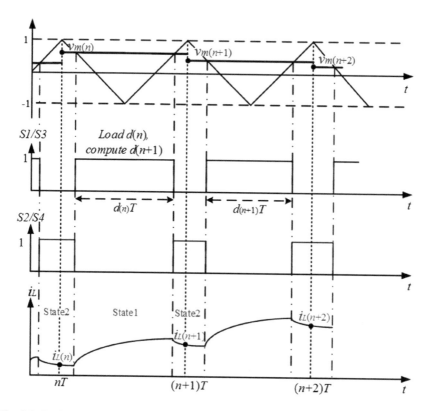

Fig. 4.2 Stroboscopic map of the single-phase PWM inverter

$$\mathbf{x} = \begin{bmatrix} i_L & v_C \end{bmatrix}^T, \quad \mathbf{A}_1 = \mathbf{A}_2 = \mathbf{A} = \begin{bmatrix} 0 & -\dfrac{1}{L} \\ \dfrac{1}{C} & -\dfrac{1}{RC} \end{bmatrix}, \quad \mathbf{B}_1 = \begin{bmatrix} \dfrac{1}{L} \\ 0 \end{bmatrix}, \quad \mathbf{B}_2 = \begin{bmatrix} -\dfrac{1}{L} \\ 0 \end{bmatrix}.$$

The duty ratio denoted as d is defined as the duration of State 1 in one switching cycle. In Fig. 4.2, it can be seen that the PWM inverter switches between State 1 and State 2 in a switching cycle, which makes the complete state equation piecewise linear and complicates the modeling process. To overcome this problem, the state-space averaging technique is employed. By averaging the durations of State 1 and State 2 in one switching cycle, a simplified state equation is derived to replace the original one, which is written as

$$\dot{\mathbf{x}} = d\left(\mathbf{A}_1\mathbf{x} + E\mathbf{B}_1\right) + \left(1-d\right)\left(\mathbf{A}_2\mathbf{x} + E\mathbf{B}_2\right) \tag{4.3}$$

namely

$$\dot{\mathbf{x}} = \mathbf{Ax} + E\mathbf{B} \tag{4.4}$$

where

$$\mathbf{B} = \begin{bmatrix} \dfrac{2d-1}{L} \\ 0 \end{bmatrix}.$$

As illustrated in Fig. 4.2, taking the filter inductor current value $i_L(n) = i_L(nT)$ and filter capacitor voltage value $v_C(n) = v_C(nT)$ at the beginning of the n-th switching cycle for initial conditions to solve Eq. (4.4), the filter inductor current value $i_L(n + 1) = i_L[(n + 1)T]$ and filter capacitor voltage value $v_C(n + 1) = v_C[(n + 1)T]$ at the beginning of $(n + 1)$-th switching cycle are derived as

$$\begin{cases} i_L(n+1) = i_L\left[(n+1)T\right] = e^{\alpha T}\left(K_1 \cos \beta T + K_2 \sin \beta T\right) - \dfrac{1}{R}\left[1 - 2d(n)\right]E \\ v_C(n+1) = v_C\left[(n+1)T\right] = e^{\alpha T}\left(K_3 \cos \beta T + K_4 \sin \beta T\right) - \left[1 - 2d(n)\right]E \end{cases} \tag{4.5}$$

where $d(n)$ is the duty ratio in the n-th switching cycle and the expressions of α, β, K_1, K_2, K_3, K_4 are denoted in Appendix A.

For the digital controller, the duty ratio d is regarded as the state variable. In terms of the discrete map in Fig. 4.2, one switching cycle delay for the digital control is taken into account, and thus the stroboscopic model of control stage is expressed as

$$d(n+1) = \frac{1}{2}v_m(n) + \frac{1}{2} \tag{4.6}$$

where

$$v_m(n) = K\left[i_C^*(n) - i_C(n)\right],$$

$$i_C^*(n) = k_p v_d^*\left(\cos \omega_f nT\right) - k_p v_C\left(nT\right) + nk_i v_d^* T\left(\cos \omega_f nT\right)$$

$$-k_i T\left(\cos \omega_f nT\right)\sum_{k=1}^{n}\left[v_C(kT)\cos\left(\omega_f kT\right) + v_C(kT - \tau)\sin\left(\omega_f kT\right)\right]$$

$$-k_i T\left(\sin \omega_f nT\right)\sum_{k=1}^{n}\left[v_C(kT)\sin\left(\omega k_f T\right) - v_C(kT - \tau)\cos\left(\omega_f kT\right)\right],$$

$$\tau = \frac{\pi}{2\omega_f}, \quad i_C(n) = i_L(n) - \frac{1}{R}v_C(n).$$

And $d(n + 1)$ is the duty ratio in the $(n + 1)$-th switching cycle. $v_m(n) = v_m(nT)$, $i_C^*(n) = i_C^*(nT)$ and $i_C(n) = i_C(nT)$ represent the modulation signal, reference filter capacitor current and filter capacitor current in the n-th switching cycle respectively. Thus, the complete stroboscopic model of the PWM inverter is described by (4.5) and (4.6). Apparently, this model is linear in one switching cycle, but on a longer time interval, it is still a nonlinear description for the PWM inverter, which can be used to investigate the dynamic properties of the system, and the accuracy of this approach will be verified by experimental results.

4.2.2   System Modeling under Inductive-Resistive Load Condition

In this case, the load Z shown in Fig. 4.1 is composed of a linear resistor of 10 Ω and a linear inductor of 4 mH, which are denoted as R_1 and L_1, respectively. Besides filter inductor current i_L and filter capacitor voltage v_C, output current i_o is also taken as a state variable for system modeling. By using the state-space averaging technique, the state equation of power stage is derived as

$$\dot{x} = A_3 x + EB_3 \tag{4.7}$$

where

$$
x = \begin{bmatrix} i_L & v_C & i_o \end{bmatrix}^T, \quad
A_3 = \begin{bmatrix} 0 & -\dfrac{1}{L} & 0 \\[6pt] \dfrac{1}{C} & 0 & -\dfrac{1}{C} \\[6pt] 0 & \dfrac{1}{L_1} & -\dfrac{R_1}{L_1} \end{bmatrix}, \quad
B_3 = \begin{bmatrix} \dfrac{2d-1}{L} \\[6pt] 0 \\[6pt] 0 \end{bmatrix}
$$

Taking the filter inductor current value $i_L(n) = i_L(nT)$, filter capacitor voltage value $v_C(n) = v_C(nT)$, and output current value $i_o(n) = i_o(nT)$ at the beginning of the n-th switching cycle for initial conditions to solve (4.7), the filter inductor current value $i_L(n + 1) = i_L[(n + 1)T]$, filter capacitor voltage value $v_C(n + 1) = v_C[(n + 1)T]$, and output current value $i_o(n + 1) = i_o[(n + 1)T)]$ at the beginning of $(n + 1)$-th switching cycle are obtained as (4.8), where the coefficients $r, \alpha_1, \beta_1, K_5, K_6, K_7$ are defined in Appendix B.

$$
\begin{cases}
i_L(n+1) = i_L\big[(n+1)T\big] = K_5 e^{rT} + e^{\alpha_1 T}\left(K_6 \cos \beta_1 T + K_7 \sin \beta_1 T\right) \\[4pt]
v_C(n+1) = v_C\big[(n+1)T\big] = -rK_5 Le^{rT} - Le^{\alpha_1 T}\left(\alpha_1 K_6 + \beta_1 K_7\right)\cos \beta_1 T - Le^{\alpha_1 T} \\[2pt]
\qquad \left(\alpha_1 K_7 - \beta_1 K_6\right)\sin \beta_1 T + \big[2d(n) - 1\big]E \\[4pt]
i_o(n+1) = i_o\big[(n+1)T\big] = \left(LCK_5 r^2 + K_5\right)e^{rT} + e^{\alpha_1 T}\left(K_6 \cos \beta_1 T + K_7 \sin \beta_1 T\right) + \\[2pt]
\qquad LC\left(\alpha_1 K_6 + \beta_1 K_7\right)e^{\alpha_1 T}\left(\alpha_1 \cos \beta_1 T - \beta_1 \sin \beta_1 T\right) \\[2pt]
\qquad + LC\left(\alpha_1 K_7 - \beta_1 K_6\right)e^{\alpha_1 T}\left(\alpha_1 \sin \beta_1 T + \beta_1 \cos \beta_1 T\right)
\end{cases}
\tag{4.8}
$$

And the modeling of the digital controller is the same as the case under resistive load condition.

4.2.3 System Modeling Under Diode Rectifier Load Condition

In this case, the structure of the studied PWM inverter is shown in Fig. 4.3.

To eliminate the odd harmonics component in the output voltage, the harmonic suppression scheme, which consists of a series of PI controllers in the SRFs with harmonic angular frequency, can be adopted. The load resistor and capacitor of the diode bridge rectifier, denoted as R_o and C_o, are 50 Ω and 2mF, respectively. An additional inductor of 2 mH is employed to form an LC filter with C_o, which is denoted as L_a. Due to the operational characteristic of diode-bridge rectifier, the on-off states of the four diodes in the rectifier are indeterminate for State 1 and State 2 during any switching cycle, which results in great difficulty in determining the output current i_o of the PWM inverter.

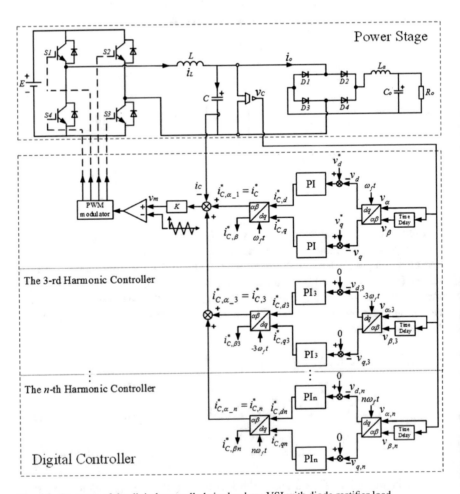

Fig. 4.3 Structure of the digital controlled single-phase VSI with diode rectifier load

In fact, accurate time-domain i_o, namely the input current of the diode bridge rectifier needed for system modeling, can be obtained only under a certain input voltage of the diode-bridge rectifier, which possesses an explicit expression on a time interval composed of one or several successive fundamental periods, rather than a switching cycle. Furthermore, the conduction angles of diode bridge rectifier are normally calculated by the iterative algorithms like Newton-Raphson method or Gauss-Seidel approach, which is complex and time-consuming. Therefore, precise modeling of the PWM inverter under diode rectifier load condition is nearly impossible with the stroboscopic map. However, by using an equivalent controlled current source (CCS) to represent the diode rectifier load, the approximate system modeling can be conducted. A simple and practical equivalent model of the diode rectifier load is proposed in [15], and described in Fig. 4.4.

Clearly, this model is an equivalent current source controlled by a dead-zone block with a proportional regulator and sinusoidal input voltage. Since the input voltage of the real diode rectifier u_{in} is quite close to the desired sinusoidal voltage $v_C^* = V_{Cm}\cos(\omega_f t)$ under the steady state of the inverter, then supposing the input voltage of the equivalent model $u_{in,e}$ is also $v_C^* = V_{Cm}\cos(\omega_f t)$, and thus parameters of the equivalent model including the regulation gain a, dead-zone limit u_L can be selected to achieve the best possible approximation of the real diode rectifier input current by simulation. Finally, the time-domain explicit expression of the PWM inverter output current i_o is approximately written as

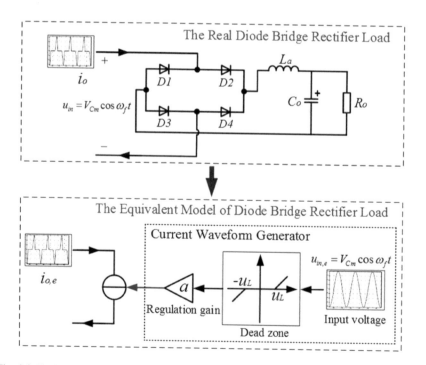

Fig. 4.4 Equivalent model of the diode bridge rectifier load

$$i_o(t) \approx i_{o,e}(t) = \begin{cases} a\left[\left(V_{Cm}\cos\omega_f t\right) - u_L\right], & V_{Cm}\cos\omega_f t > u_L \\ 0, & -u_L \leq V_{Cm}\cos\omega_f t \leq u_L \quad (4.9) \\ a\left[-u_L - \left(V_{Cm}\cos\omega_f t\right)\right], & V_{Cm}\cos\omega_f t < -u_L \end{cases}$$

Once again, by taking the filter inductor current i_L and filter capacitor voltage v_C as the state variables, the simplified state equation of power stage is derived as

$$\dot{\mathbf{x}} = \mathbf{A_4}\mathbf{x} + \mathbf{B_4} \tag{4.10}$$

where

$$\mathbf{x} = \begin{bmatrix} i_L & v_C \end{bmatrix}^T, \quad \mathbf{A_4} = \begin{bmatrix} 0 & -\dfrac{1}{L} \\ \dfrac{1}{C} & 0 \end{bmatrix}, \quad \mathbf{B_4} = \begin{bmatrix} \dfrac{(2d-1)E}{L} \\ -\dfrac{1}{C}io \end{bmatrix}$$

Mathematically, taking the filter inductor current value $i_L(n) = i_L(nT)$ and filter capacitor voltage value $v_C(n) = v_C(nT)$ at the beginning of the n-th switching cycle for the initial conditions to solve (4.10), the filter inductor current value $i_L(n + 1) = i_L[(n + 1)T]$ and the filter capacitor voltage value $v_C(n + 1) = v_C[(n + 1)T]$ can be derived.

For the digital controller, $v_m(n)$ is modified as

$$v_m(n) = \left[K\sum_{m=0}^{n} i^*_{C,\alpha\_2m+1}(n) \right] - Ki_C(n) \tag{4.11}$$

The output of the $(2m + 1)$-th order harmonic controller is represented as (4.12), where $\tau = \pi/2\omega_f$, $i_C(n) = i_L(n)-i_o(n)$.

$$i^*_{C,\alpha\_2m+1}(n) = -k_{p,2m+1}v_C(nT) - k_{i,2m+1}T\cos$$

$$\left[(2m+1)\omega_f nT\right]\sum_{k=1}^{n}\begin{Bmatrix} v_C(kT)\cos\left[(2m+1)\omega_f kT\right] + \\ v_C(kT-\tau)\sin\left[(2m+1)\omega_f kT\right] \end{Bmatrix}$$

$$-k_{i,2m+1}T\sin\left[(2m+1)\omega_f nT\right]\sum_{k=1}^{n}\begin{Bmatrix} v_C(kT)\sin\left[(2m+1)\omega_f kT\right] - \\ v_C(kT-\tau)\cos\left[(2m+1)\omega_f kT\right) \end{Bmatrix} \tag{4.12}$$

The parameters $k_{p,2m+1}$ and $k_{i,2m+1}$ are the proportional gain and integral gain of the $(2m + 1)$-th harmonic controller, and $d(n + 1)$ keeps the same form as the case under resistive load condition.

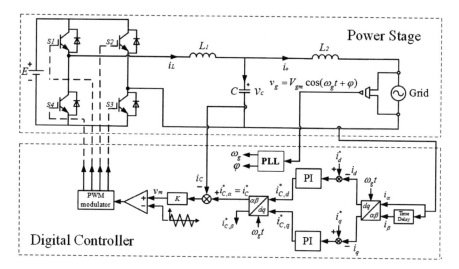

Fig. 4.5 System structure of the digital controlled single-phase VSI with SRF control loop in grid-connected mode

4.2.4 System Modeling of the PWM Inverter in the Grid-Connected Mode

In the grid-connected mode, an LCL filter is employed to replace the LC filter used in stand-alone mode, as presented in Fig. 4.5.

The inverter output current i_o must be controlled to gain the same frequency as the grid voltage, which is also mainly implemented in the SRF. Supposing the grid voltage is $v_g = V_{gm}\cos(\omega_g t + \varphi)$, the reference of i_o is accordingly defined as $i_o^* = I_{om}\cos(\omega_g t + \varphi)$. As a result, the d-axis reference current i_d^* is determined as $I_{om}\cos\varphi$, and q-axis reference current i_q^* is set as $I_{om}\sin\varphi$. The angular frequency ω_g and phase φ of the grid voltage are identified by a single-phase phase-locked loop (PLL), and normally, ω_g and φ are constant for a definite single-phase grid. Taking filter inductor current i_L, output current i_o, filter capacitor v_C as state variables, and following the aforementioned modeling steps, the stroboscopic model of the PWM inverter in grid-connected mode can also be easily established.

4.3 Stability Analysis Under Parameter Variations in Voltage Loop

Since the fictitious second phase voltage v_β is generated by delaying v_C for one quarter of the fundamental cycle of v_C^*, it is extremely difficult to obtain the exact values of v_β at the beginning of any switching cycle for the direct iteration defined by the stroboscopic model. Thus, the Jacobian matrix method and the Lyapunov exponent method are employed for the stability analysis of the single-phase PWM inverter.

4.3.1 Stability Analysis Under Resistive Load by Using Jacobian Matrix Method

Denoting the fixed point of the stroboscopic model as $\left(i_L^*, v_C^*, d^*\right)$ and substituting it into (4.5) and (4.6), and supposing $i_L(n+1) = i_L(n) = i_L^*$, $v_C(n+1) = v_C(n) = v_C^*$, $d(n+1) = d(n) = d^*$, the Jacobian matrix at the fixed point is derived as (4.13), where the matrix elements are listed in Appendix C.

$$
\mathbf{J} = \begin{bmatrix}
\dfrac{\partial i_L(n+1)}{\partial i_L(n)} & \dfrac{\partial i_L(n+1)}{\partial v_C(n)} & \dfrac{\partial i_L(n+1)}{\partial d(n)} \\[2mm]
\dfrac{\partial v_C(n+1)}{\partial i_L(n)} & \dfrac{\partial v_C(n+1)}{\partial v_C(n)} & \dfrac{\partial v_C(n+1)}{\partial d(n)} \\[2mm]
\dfrac{\partial d(n+1)}{\partial i_L(n)} & \dfrac{\partial d(n+1)}{\partial v_C(n)} & \dfrac{\partial d(n+1)}{\partial d(n)}
\end{bmatrix}_{\left(i_L^*, v_C^*, d^*\right)}
\tag{4.13}
$$

As can be seen, the Jacobian matrix in (4.13) is independent to any fixed point of the stroboscopic model, and this is because the stabilities of different fixed points are regarded as consistent for the established state-space averaging model.

To investigate the fast-scale stability of the PWM inverter with a low computation cost, four typical values of k_i, including 20, 40, 60, 80, are taken into account to find the stability regions by the Jacobian matrix method, which are depicted in Fig. 4.6.

In Fig. 4.6, the green zones represent the control parameters that enable all eigenvalues of the Jacobian matrix in (4.13) to be located in the unit circle on the complex plane, which means that the PWM inverter is stable. However, the red zones represent parameters that make at least one eigenvalue of the Jacobian matrix lies on or outside the unit circle, which indicates that the PWM inverter is unstable.

It is obvious that, for a certain k_i, stable intervals of K become smaller as k_p increases. Conversely, when K increases, stable intervals of k_p reduce. However, the effect of parameter k_i on the stability of the PWM inverter is not distinct. This is because the Jacobian matrix in (4.13) contains only one component of k_i, which is $(1/2)KTk_i$. Owing to a sufficiently small switching cycle T, it is hard for $(1/2)KTk_i$ to exert a significant influence on the eigenvalues of the Jacobian matrix.

In consideration of the stability regions in Fig. 4.6 and without loss of generality, $k_i = 20$ and $K = 0.5$ is taken as a typical condition to analyze the effect of k_p on the fast-scale stability of the PWM inverter in detail. Denoting three eigenvalues of the

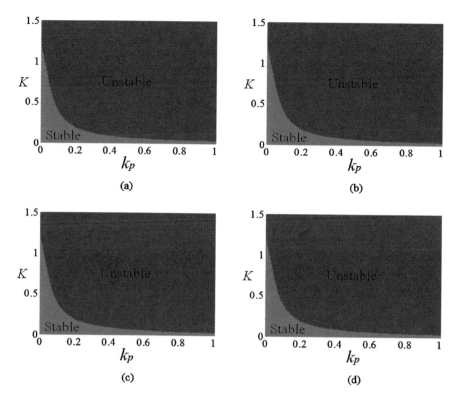

Fig. 4.6 Stability regions of the single-phase VSI with SRF voltage control under different k_i. (**a**) $k_i = 20$; (**b**) $k_i = 40$; (**c**) $k_i = 60$; (**d**) $k_i = 80$

Jacobian matrix in (4.13) as λ_1, λ_2, λ_3, and then their loci and moduli are illustrated in Fig. 4.7.

As shown in Fig. 4.7, λ_1 and λ_2 form a pair of complex-conjugates, and λ_3 remains real on the studied interval consistently. When $0 < k_p < 0.082$, λ_1, λ_2, and λ_3 all lie in the unit circle, which suggests that the PWM inverter is stable. When $0.082 < k_p < 1$, λ_1 and λ_2 move outside the unit circle, while λ_3 still locates in it, which indicates that the PWM inverter becomes unstable. Thus, $k_p = 0.082$ is the critical point determining the stable and unstable state of the PWM inverter when $k_i = 20$ and $K = 0.5$.

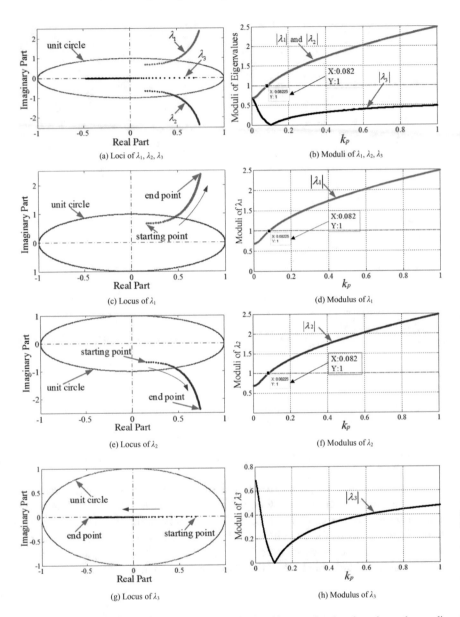

Fig. 4.7 Loci and moduli of three eigenvalues of the Jacobian matrix when k_p varies under condition of $k_i = 20$ and $K = 0.5$. (**a**) Loci of $\lambda_1, \lambda_2, \lambda_3$, (**b**) Moduli of $\lambda_1, \lambda_2, \lambda_3$, (**c**) Locus of λ_1, (**d**) Modulus of λ_1, (**e**) Locus of λ_2, (**f**) Modulus of λ_2, (**g**) Locus of λ_3, (**h**) Modulus of λ_3

4.3.2 Stability Analysis Under Resistive Load Condition by Using Lyapunov Exponent Method

To validate the above results obtained by the Jacobian matrix method, the maximum Lyapunov exponents of the inverter are calculated to show its slow-scale stability. Referring to [9], the maximum Lyapunov exponent of a three-dimensional discrete system can be defined as

$$\lambda L = \max \left(\lambda_{L1}, \lambda_{L2}, \lambda_{L3} \right) \tag{4.14}$$

$$\begin{bmatrix} \lambda_{L1} \\ \lambda_{L2} \\ \lambda_{L3} \end{bmatrix} = \lim_{n \to \infty} \frac{1}{n} \ln \left\| \left[eig\left(JnJn - 1 \cdots J1 \right) \right] \right\| \tag{4.15}$$

where $eig(J_n J_{n-1} \ldots J_1)$ is the eigenvalue function of $J_n J_{n-1} \ldots J_1$, and J_n is the Jacobian matrix at the mapping point in the n-th switching cycle. Moreover, in terms of the stroboscopic model in (4.5) to (4.6), it is possible to derive $J_1 = J_2 = \ldots J_{n-1} = J_n = J$, in which J is the Jacobian matrix presented in (4.13).

Figure 4.8 illustrates the projections of maximum Lyapunov exponent spectrums on K-k_p plane, under $k_i = 20, 40, 60, 80$. The red regions represent the control parameters leading to positive or zero maximum Lyapunov exponent, which are defined as unstable regions for the inverter. And the green regions match the control parameters producing negative maximum Lyapunov exponent, which are accordingly defined as stable regions. Clearly, Fig. 4.8 is almost the same as Fig. 4.6, and that means the slow-scale analysis results obtained by Lyapunov exponent method are consistent with the fast-scale analysis results obtained by the Jacobian matrix method.

In fact, since $J_1 = J_2 = \ldots J_{n-1} = J_n = J$, it is possible to obtain $eig(J_n J_{n-1} \ldots J_1) = [eig(J)]^n$ according to matrix theory, which yields

$$\begin{bmatrix} \lambda_{L1} \\ \lambda_{L2} \\ \lambda_{L3} \end{bmatrix} = \lim_{n \to \infty} \frac{1}{n} \ln \left\| \left[eig\left(JnJn - 1 \cdots J1 \right) \right] \right\|$$

$$= \lim_{n \to \infty} \frac{1}{n} \ln \left\| \left[eig\left(J \right) \right]^n \right\| \tag{4.16}$$

$$= \ln \left\| \left[eig\left(J \right) \right] \right\|$$

Obviously, Eq. (4.16) shows a direct proof for the equivalence of Jacobian matrix method and Lyapunov exponent method, which also reveals that, in the sense of state space averaging, the fast-scale stability and slow-scale stability are almost the same for the studied PWM inverter model.

The maximum Lyapunov exponent spectrum on k_p is presented in Fig. 4.9 under condition of $k_i = 20$ and $K = 0.5$. As can be seen, when $k_p < 0.082$, the maximum

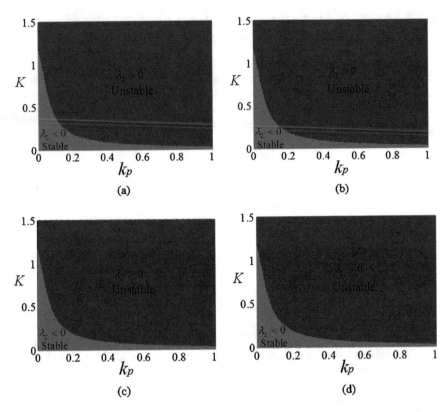

Fig. 4.8 Projections on the K-k_p plane of maximum Lyapunov exponent spectrums under different k_i. (a) $k_i = 20$; (b) $k_i = 40$; (c) $k_i = 60$; (d) $k_i = 80$

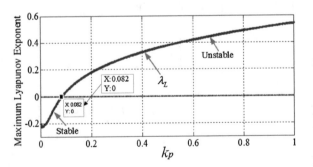

Fig. 4.9 The maximum Lyapunov exponent spectrum on k_p, under resistive load condition when $k_i = 20$ and $K = 0.5$

Lyapunov exponent is negative, but when $k_p > 0.082$, the maximum Lyapunov exponent becomes positive, which means that $k_p = 0.082$ is the critical point when $k_i = 20$ and $K = 0.5$. Figure 4.9 clearly shows a good consistency with Fig. 4.7.

4.3.3   Stability Analysis Under Inductive-Resistive Load Condition

In view of the consistency of the Jacobian matrix method and the Lyapunov exponent method, proved in the analyses under resistive load condition, and the significant similarities between the system models of resistive and inductive-resistive load, it is reasonable to infer that Jacobian matrix method and Lyapunov exponent method are also coincident for inductive-resistive load, which reveals the fact that inductive-resistive load is inherently a kind of linear load. Therefore, for the sake of brevity, only Lyapunov exponent method is adopted to investigate the effect of k_p on the stability of the PWM inverter when $k_i = 20$ and $K = 0.5$ in this part. The result is presented in Fig. 4.10. It is clear that, when $k_p < 0.07$, the maximum Lyapunov exponent is negative, but when $k_p > 0.07$, the maximum Lyapunov exponent becomes positive. So $k_p = 0.07$ is the critical point for system stability when $k_i = 20$ and $K = 0.5$ under inductive-resistive load condition.

4.3.4   Stability Analysis Under Nonlinear Load Condition

Since the equivalent controlled current source is an approximated time-domain model of the diode rectifier with limited precision, it is not very suitable for fast-scale stability analysis of the PWM inverter which is sensitive to the model accuracy. Hence, in this part, only slow-scale stability analysis under diode rectifier load condition is conducted when $k_i = 20$ and $K = 0.5$ by employing the equivalent diode rectifier model and Lyapunov exponent method. The 3rd, 5th, 7th, 9th, and 11th harmonic control schemes are also added into the controller, and the parameters of them are selected by simulation method and considered as constants in the analysis. The result is presented in Fig. 4.11. It can be seen that, when $k_p < 0.048$, the maximum Lyapunov exponent is negative, but when $k_p > 0.048$, the maximum Lyapunov exponent becomes positive, which means that $k_p = 0.048$ is the critical point when

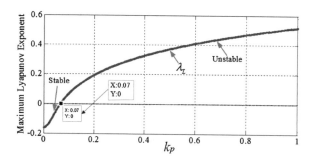

Fig. 4.10 The maximum Lyapunov exponent spectrum on k_p under inductive-resistive load condition when $k_i = 20$ and $K = 0.5$

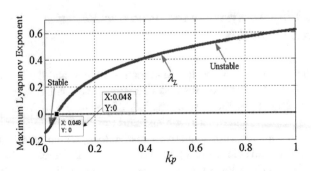

Fig. 4.11 The maximum Lyapunov exponent spectrum on k_p under nonlinear load condition when $k_i = 20$ and $K = 0.5$

$k_i = 20$ and $K = 0.5$ under nonlinear load condition. Apparently, the critical point of k_p is smaller under nonlinear load condition, compared to the cases of linear load conditions with same k_i and K.

4.4 Stability Analysis Under Parameter Variations in Current Loop

4.4.1 Stability Analysis Under Resistive Load Condition by Using Jacobian Matrix Method

Considering the analysis in the previous section, the effect of K on the stability of the PWM inverter is investigated under condition of $k_i = 20$ and $k_p = 0.04$ by the Jacobian matrix method first. Figure 4.12 shows the loci and moduli of the three eigenvalues of the Jacobian matrix in (4.13), under variations of parameter K.

As shown in Fig. 4.12, on the studied interval of K, λ_1 and λ_2 are a pair of complex-conjugates. λ_3 is real and remains in the unit circle when K varies. When $K < 0.742$, λ_1 and λ_2 are located in the unit circle, but when $K > 0.742$, they move outside the unit circle. Hence the PWM inverter is stable when $K < 0.742$, but unstable when $K > 0.742$, and $K = 0.742$ is the critical point for the stability of the inverter when $k_i = 20$ and $k_p = 0.04$.

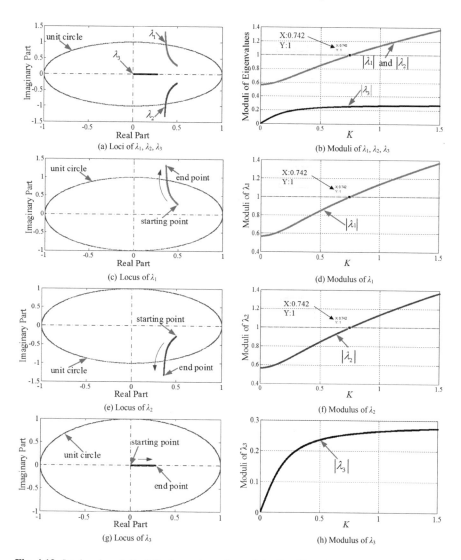

Fig. 4.12 Loci and moduli of the three eigenvalues of the Jacobian matrix when K varies under condition of $k_i = 20$ and $k_p = 0.04$. (**a**) Loci of $\lambda_1, \lambda_2, \lambda_3$, (**b**) Moduli of $\lambda_1, \lambda_2, \lambda_3$, (**c**) Locus of λ_1, (**d**) Modulus of λ_1, (**e**) Locus of λ_2, (**f**) Modulus of λ_2, (**g**) Locus of λ_3, (**h**) Modulus of λ_3

4.4.2 Stability Analysis Under Resistive Load Condition by Using Lyapunov Exponent Method

Figure 4.13 depicts the maximum Lyapunov exponent spectrum on K when $k_i = 20$ and $k_p = 0.04$ to validate the results in Fig. 4.12. In Fig. 4.13, it is shown that when $K < 0.742$, the maximum Lyapunov exponent is negative. However, when $K > 0.742$,

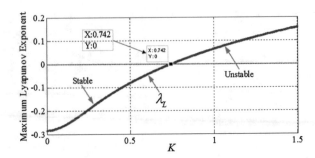

Fig. 4.13 The maximum Lyapunov exponent spectrum on K under resistive load condition when $k_i = 20$ and $k_p = 0.04$

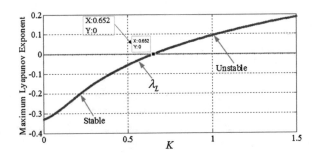

Fig. 4.14 The maximum Lyapunov exponent spectrum on K under inductive-resistive load condition when $k_i = 20$ and $k_p = 0.04$

the maximum Lyapunov exponent becomes positive, which means that $K = 0.742$ is the critical point. Obviously, Fig. 4.13 shows a good consistency with Fig. 4.12.

4.4.3   Stability Analysis Under Inductive-Resistive Load Condition

Under inductive-resistive load condition, Lyapunov exponent method is adopted to investigate the effect of K on the stability of the PWM inverter when $k_i = 20$ and $k_p = 0.04$, and the result is shown as Fig. 4.14. It can be seen from Fig. 4.14, when $K < 0.652$, the maximum Lyapunov exponent is negative, but when $K > 0.652$, the maximum Lyapunov exponent becomes positive. So, $K = 0.652$ is the critical point when $k_i = 20$ and $k_p = 0.04$ under inductive-resistive load condition.

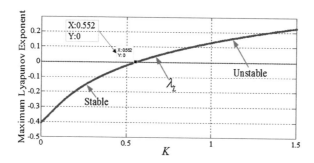

Fig. 4.15 The maximum Lyapunov exponent spectrum on K under nonlinear load condition when $k_i = 20$ and $k_p = 0.04$

4.4.4   Stability Analysis Under Nonlinear Load Condition

Under nonlinear load condition, slow-scale analysis is done to investigate the effect of K on the stability of the PWM inverter when $k_i = 20$ and $k_p = 0.04$, based on the equivalent model of diode rectifier and Lyapunov exponent method. The 3rd, 5th, 7th, 9th, and 11th harmonic controllers are also employed in this case, and the parameters of them keep the same as those in the analysis on k_p in the previous section. The analysis result is shown in Fig. 4.15. It is clear that, when $K < 0.552$, the maximum Lyapunov exponent is negative. However, when $K > 0.552$, the maximum Lyapunov exponent becomes positive, and that means $K = 0.552$ is the critical point under nonlinear load when $k_i = 20$ and $k_p = 0.04$. Evidently, this critical point of K is obviously much smaller than those of linear load conditions with same k_i and k_p.

4.5   Simulation Results and Discussions

This section presents the simulation results obtained from EMTP software, as a supplement for the previous analytical results, where the effects of digital control delay, load parameters, and controller gain on closed-loop stability boundaries are considered. Moreover, the approximation model of control delay using the transport delay model, first-order low-pass filter, and the lead-lag compensator models are studied. The transport delay model is found to be more accurate, compared with the analytical and experimental results. However, the first-order low-pass filter and lead-lag compensator models provide more optimistic evaluations for the stability boundaries, that is, an excessive higher controller gain might be deduced, which may violate the closed-loop stability criterion for the practical systems.

4.5.1   Stability Evaluation with a Control Delay of 50 Microsecond

Figures 4.16, 4.17, and 4.18 show the simulation results when $k_p = 0.082$, $k_i = 20$, $R = 20$ Ohm, and the effective control delay is $T_d = 50$e-6 µs. Figure 4.16 shows the case when the control delay is denoted using the transport delay model. It can be observed from Fig.4.16 that when the inner loop controller gain K is less than 0.7, the closed-loop stability is ensured. When the inner loop controller gain K exceeds 0.706, obvious oscillation instability would occur due to insufficient stability margin. Figure 4.17 shows the case when the control delay is denoted using the first-order low-pass filter model. It can be observed from Fig.4.17 that when the inner loop controller gain K is less than 4.5, the closed-loop stability is ensured. When the inner loop controller gain K exceeds 4.7, obvious oscillation instability would occur due to insufficient stability margin. Figure 4.18 shows the case when the control delay is denoted using the lead-lag compensator model. It can be observed from Fig.4.18 that when the inner loop controller gain K is less than 6.3, the closed-loop stability is ensured. When the inner loop controller gain K exceeds 6.4, obvious oscillation instability would occur due to insufficient stability margin.

The stability boundary obtained from the transport delay model is much smaller than the cases when the first-order low-pass filter or the lead-lag compensator model is utilized to approximate the effect of control delay. Next, the control delay is extended to 75 and 100 microseconds to study the stability boundaries under different control delay models.

4.5.2   Stability Evaluation with a Control Delay of 75 Microseconds

Figures 4.19, 4.20, and 4.21 show the simulation results when $k_p = 0.082$, $k_i = 20$, $R = 20$ Ohm, and the effective control delay is $T_d = 75$e-6 µs. Figure 4.19 shows the case when the control delay is denoted using transport delay model. It can be observed from Fig. 4.19 that when the inner loop controller gain K is less than 0.475, the closed-loop stability is ensured. When the inner loop controller gain K exceeds 0.48, obvious oscillation instability would occur due to insufficient stability margin. Figure 4.20 shows the case when the control delay is denoted using the first-order low-pass filter model. It can be observed from Fig. 4.20 that when the inner loop controller gain K is less than 3.3, the closed-loop stability is ensured. When the inner loop controller gain K exceeds 3.4, obvious oscillation instability would occur due to insufficient stability margin. Figure 4.21 shows the case when the control delay is denoted using the lead-lag compensator model. It can be observed from Fig. 4.21 that when the inner loop controller gain K is less than 4.3, the closed-loop stability is ensured. When the inner loop controller gain K exceeds 4.4, obvious oscillation instability would occur due to insufficient stability margin.

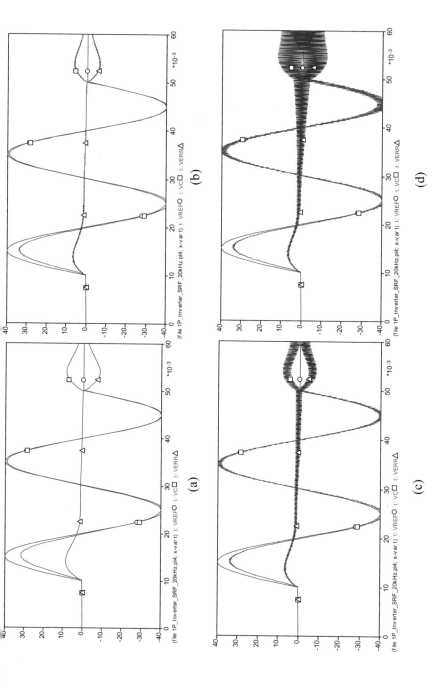

Fig. 4.16 Simulation results when $k_p = 0.082$, $k_i = 20$, $R = 20$ Ohm, and the control delay is denoted using transport delay $T_d = 50e\text{-}6$ μs. (**a**) $K = 0.5$; (**b**) $K = 0.7$; (**c**) $K = 0.706$; (**d**) $K = 0.708$

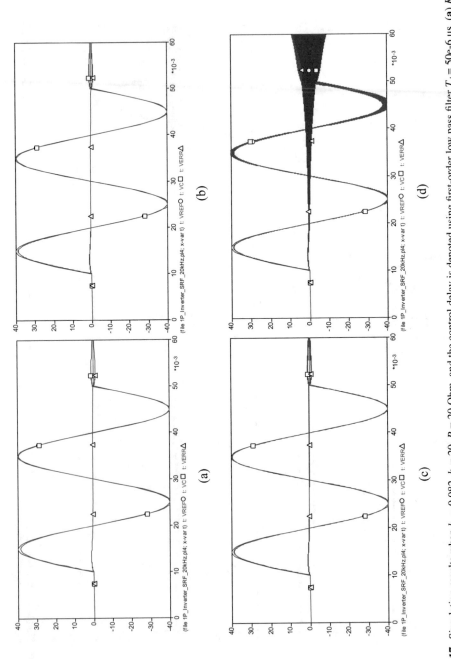

Fig. 4.17 Simulation results when $k_p = 0.082$, $k_i = 20$, $R = 20$ Ohm, and the control delay is denoted using first-order low pass filter $T_d = 50e\text{-}6$ μs. (**a**) $K = 4$; (**b**) $K = 4.5$; (**c**) $K = 4.6$; (**d**) $K = 4.7$

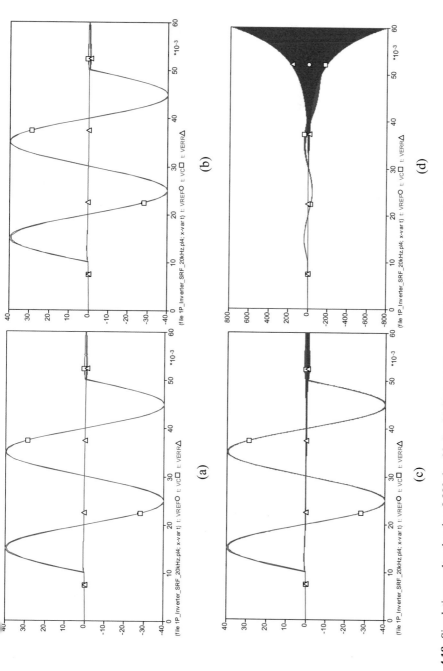

Fig. 4.18 Simulation results when $k_p = 0.082$, $k_i = 20$, $R = 20$ Ohm, and the control delay is denoted using lead-lag compensator $T_d = 50e$-6 μs. (**a**) $K = 6.2$; (**b**) $K = 6.3$; (**c**) $K = 6.4$; (**d**) $K = 6.5$

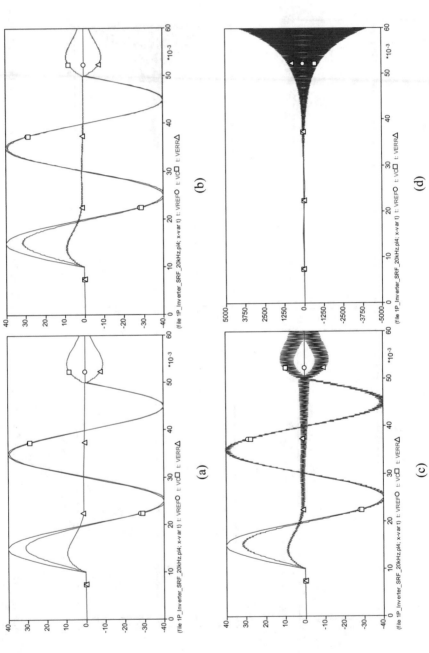

Fig. 4.19 Simulation results when $k_p = 0.082$, $k_i = 20$, $R = 20$ Ohm, and the control delay is denoted using transport delay $T_d = 75\text{e-}6$ μs. (**a**) $K = 0.45$; (**b**)

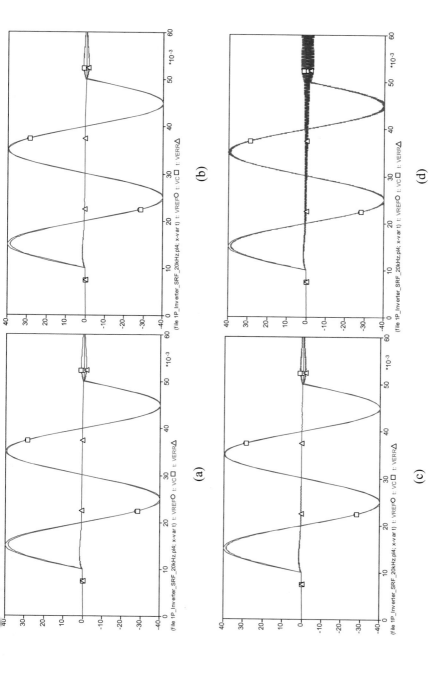

Fig. 4.20 Simulation results when $k_p = 0.082$, $k_i = 20$, $R = 20$ Ohm, and the control delay is denoted using first-order low pass filter $T_d = 75\text{e-}6$ µs. (**a**) $K = 3.2$; (**b**) $K = 3.3$; (**c**) $K = 3.4$; (**d**) $K = 3.5$

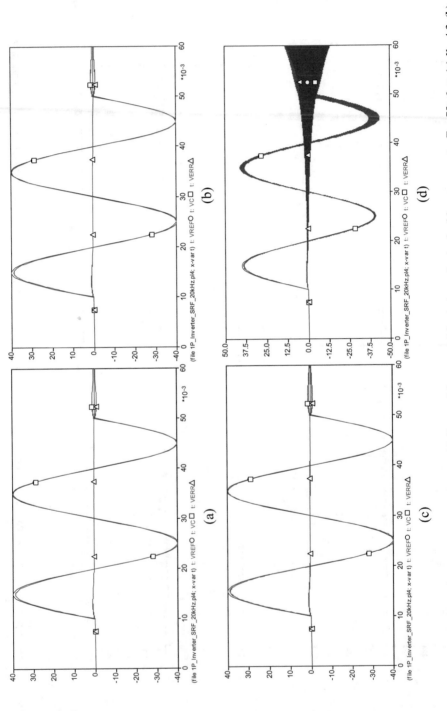

Fig. 4.21 Simulation results when $k_p = 0.082$, $k_i = 20$, $R = 20$ Ohm, and the control delay is denoted using lead-lag compensator $T_d = 75e\text{-}6$ μs. (**a**) $K = 4.2$; (**b**) $K = 4.3$; (**c**) $K = 4.4$; (**d**) $K = 4.5$

4.5.3 Stability Evaluation with a Control Delay of 100 Microseconds

Figures 4.22, 4.23, and 4.24 show the simulation results when $k_p = 0.082$, $k_i = 20$, $R = 20$ Ohm, and the effective control delay is $T_d = 100\text{e-}6$ µs. Figure 4.22 shows the case when the control delay is denoted using the transport delay model. It can be observed from Fig. 4.22 that when the inner loop controller gain K is less than 0.38, the closed-loop stability is ensured. When the inner loop controller gain K exceeds 0.385, obvious oscillation instability would occur due to insufficient stability margin. Figure 4.23 shows the case when the control delay is denoted using the first-order low-pass filter model. It can be observed from Fig. 4.23 that when the inner loop controller gain K is less than 3.1, the closed-loop stability is ensured. When the inner loop controller gain K exceeds 3.2, obvious oscillation instability would occur due to insufficient stability margin. Figure 4.24 shows the case when the control delay is denoted using the lead-lag compensator model. It can be observed from Fig. 4.24 that when the inner loop controller gain K is less than 3.4, the closed-loop stability is ensured. When the inner loop controller gain K exceeds 3.5, obvious oscillation instability would occur due to insufficient stability margin.

4.5.4 Stability Evaluation with the Variation of Load Parameters

Next, the effects of load parameters are evaluated in terms of closed-loop stability in the analytical model. Figure 4.25 shows the simulation results when $k_p = 0.082$, $k_i = 20$, with the RL load: $R = 20$ Ohm, $L_{load} = 0.5$ mH, and the control delay is denoted using transport delay $T_d = 75\text{e-}6$ µs. It shows that when the inner loop gain K is less than 0.35, the closed-loop stability is ensured. However, when the inner loop gain K is greater than 0.35, the system instability would occur due to insufficient stability margin.

Figure 4.26 shows the simulation results when $k_p = 0.082$, $k_i = 20$, with the RL load: $R = 5$ Ohm, $L_{load} = 0.5$ mH, and the control delay is denoted using transport delay $T_d = 75\text{e-}6$ µs. It shows that when the inner loop gain K is less than 0.14, the closed-loop stability is ensured. However, when the inner loop gain K is greater than 0.15, the system instability would occur due to insufficient stability margin. As a comparison to the case in Fig.4.25, it can be inferred that, under heavy load conditions, the inner-loop gain should be reduced, in order to achieve a tradeoff between the dynamic response and closed-loop stability.

On the other hand, Fig. 4.27 shows the simulation results when $k_p = 0.082$, $k_i = 20$, with the RL load: $R = 20$ Ohm, $L_{load} = 1$ mH, and the control delay is denoted using transport delay $T_d = 75\text{e-}6$ µs. It shows that when the inner loop gain K is less than 0.18, the closed-loop stability is ensured. However, when the inner loop gain K is greater than 0.19, the system instability would occur due to insufficient stability

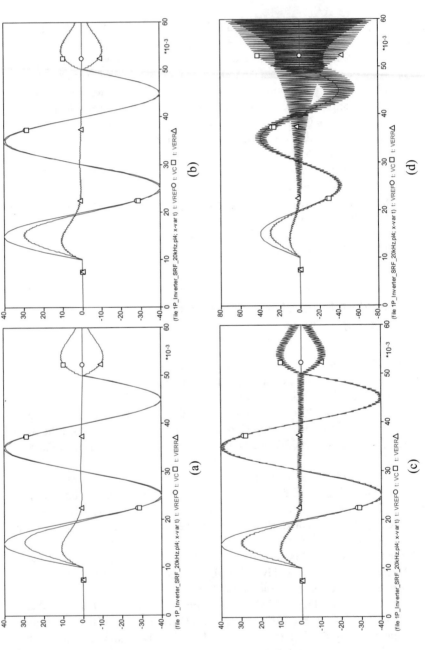

Fig. 4.22 Simulation results when $k_p = 0.082$, $k_i = 20$, $R = 20$ Ohm, and the control delay is denoted using transport delay $T_d = 100e\text{-}6$ μs. (**a**) $K = 0.37$; (**b**) $K = 0.28$; (**c**) $K = 0.385$; (**d**) $K = 0.39$

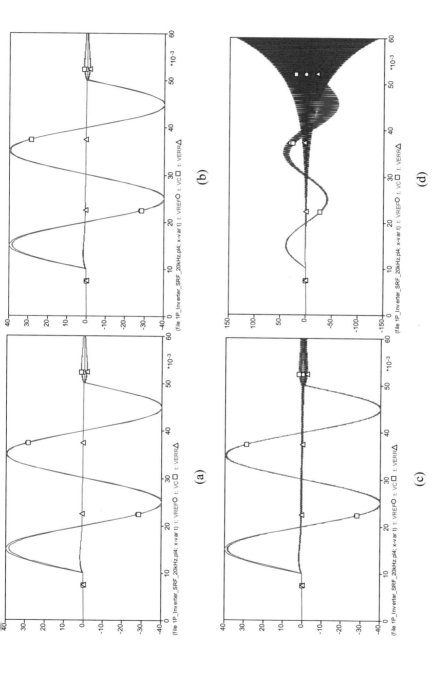

Fig. 4.23 Simulation results when $k_p = 0.082$, $k_i = 20$, $R = 20$ Ohm, and the control delay is denoted using first-order low pass filter $T_d = 100\text{e-}6$ μs. (**a**) $K = 3$; (**b**) $K = 3.1$; (**c**) $K = 3.2$; (**d**) $K = 3.3$

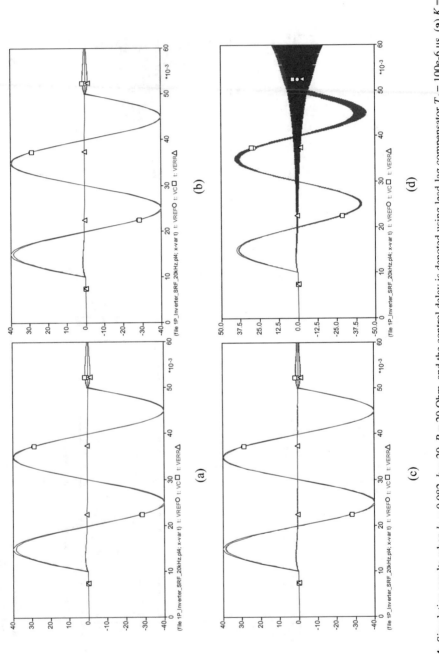

Fig. 4.24 Simulation results when $k_p = 0.082$, $k_i = 20$, $R = 20$ Ohm, and the control delay is denoted using lead-lag compensator $T_d = 100e\text{-}6$ μs. (**a**) $K = 3.3$; (**b**) $K = 3.4$; (**c**) $K = 3.5$; (**d**) $K = 3.6$

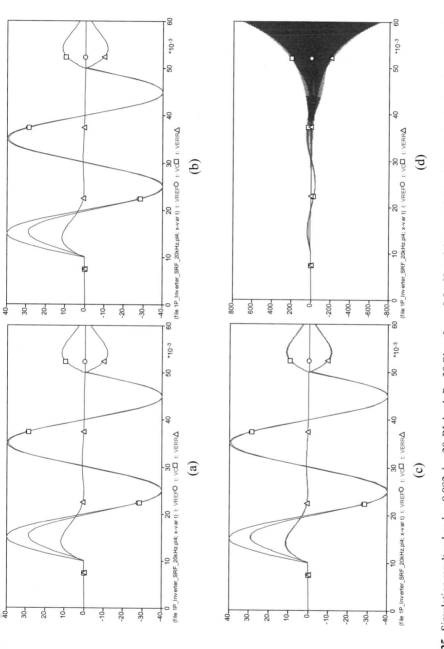

Fig. 4.25 Simulation results when $k_p = 0.082$, $k_i = 20$, RL load: $R = 20$ Ohm, $L_{load} = 0.5$ mH, and the control delay is denoted using transport delay $T_d = 75$e-6 μs. (**a**) $K = 0.33$; (**b**) $K = 0.34$; (**c**) $K = 0.35$; (**d**) $K = 0.36$

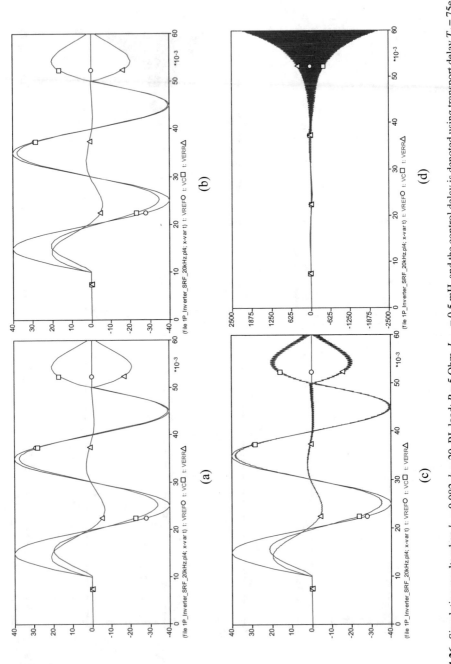

Fig. 4.26 Simulation results when $k_p = 0.082$, $k_i = 20$, RL load: $R = 5$ Ohm, $L_{load} = 0.5$ mH, and the control delay is denoted using transport delay $T_d = 75e\text{-}6$ μs. (a) $K = 0.13$; (b) $K = 0.14$; (c) $K = 0.15$; (d) $K = 0.16$

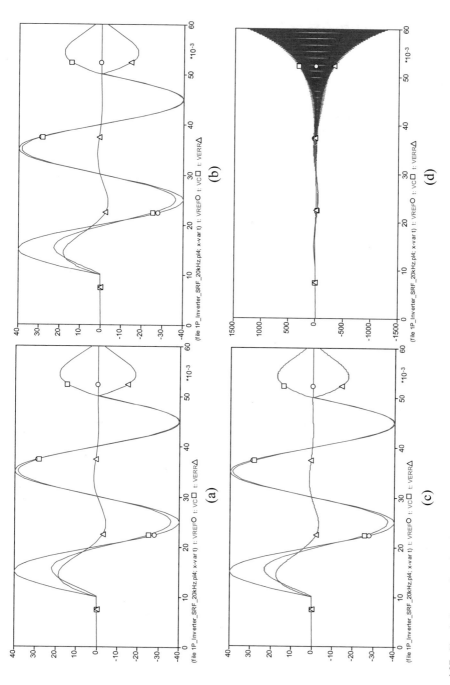

Fig. 4.27 Simulation results when $k_p = 0.082$, $k_i = 20$, RL load: $R = 20$ Ohm, $L_{load} = 1$ mH, and the control delay is denoted using transport delay $T_d = 75e\text{-}6$ µs. (**a**) $K = 0.16$; (**b**) $K = 0.17$; (**c**) $K = 0.18$; (**d**) $K = 0.19$

margin. Similarly, as a comparison to the case in Fig. 4.25, it can be inferred that when the load is more inductive, the inner-loop gain should be reduced, in order to achieve a tradeoff between the dynamic response and closed-loop stability.

4.6   Experimental Results

According to the system structure in Fig. 4.1, an experimental setup is built with a controller of TMS320F28335 DSP to verify the theoretical analyses. Voltage sensor HPT205A and current sensor ACS712ELCTR-05B-T are employed. The transformation ratio of HPT205A is 2 mA: 2 mA, and its precision is 0.1%. The optimized range of ACS712ELCTR-05B-T is ±5A, and its sensitivity is 185 mV/A. The DC-link voltage of the inverter is provided by a programmable DC power supply. The RIGOL digital oscilloscope is employed to record the time-domain waveforms and Fast Fourier Transform (FFT) results. The experimental results are presented as follows.

4.6.1   Experimental Results Under Resistive Load Condition

Figure 4.28 shows the steady-state waveforms under resistive load condition for different k_p when $k_i = 20$, $K = 0.5$. It is evident that waveforms of the filter capacitor voltage v_C and output current i_o are periodic and sinusoidal without any distortion when $k_p = 0.042$, which indicates that the PWM inverter is stable. When $k_p = 0.062$, waveforms of v_C and i_o become slightly distorted, which means that the PWM inverter is nearly critical stable. When k_p increases to 0.082, waveforms of v_C and i_o are obviously distorted, which suggests that the PWM inverter is oscillating.

And when $k_p = 0.102$, serious oscillation appears in the waveforms of v_C and i_o, and the PWM inverter is totally unstable. Therefore, experimental results in Fig. 4.28 are in accordance with the theoretical results that the PWM inverter becomes unstable when $k_p > 0.082$ for $k_i = 20$ and $K = 0.5$. Besides the waveforms observed in several fundamental cycles of v_C, which describe the slow-scale dynamic behaviors of the inverter, the magnified waveforms of v_C and i_o in successive switching cycles are also provided to present the fast-scale dynamic behaviors of the inverter.

Obviously, the magnified waveforms demonstrate almost the same stability characteristics as the normal waveforms, and that means the fast- and slow-scale stabilities are consistent for the inverter under resistive load condition, which is also consistent with the theoretical results. The transient waveforms under resistive load condition when $k_i = 20$, $K = 0.5$, and $k_p = 0.04$ are presented in Fig. 4.29, including the transient waveforms in response to no load to nominal resistive load step change, and transient waveforms in response to +50% step change of load resistor. Clearly, the transient waveforms prove that the dynamic response of the PWM inverter with resistive load is quite fast.

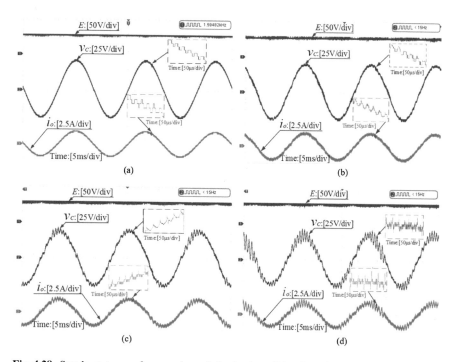

Fig. 4.28 Steady-state waveforms under resistive load condition for different k_p when $k_i = 20$ and $K = 0.5$. (**a**) $k_p = 0.042$; (**b**) $k_p = 0.062$; (**c**) $k_p = 0.082$; (**d**) $k_p = 0.102$

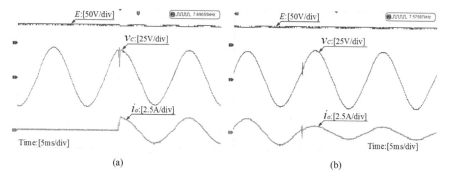

Fig. 4.29 Transient waveforms under resistive load condition when $k_i = 20$, $K = 0.5$ and $k_p = 0.04$ (**a**) Transient waveforms in response to no load to nominal resistive load step change; (**b**) Transient waveforms in response to +50% step change of load resistor

Figure 4.30 illustrates the steady-state waveforms under resistive load condition for different K when $k_i = 20$ and $k_p = 0.04$. As shown in Fig. 4.30, waveforms of the filter capacitor voltage v_C and output current i_o are periodic and sinusoidal without any distortion when $K = 0.542$, which indicates that the PWM inverter is stable. When $K = 0.642$, waveforms of v_C and i_o become slightly distorted, which means

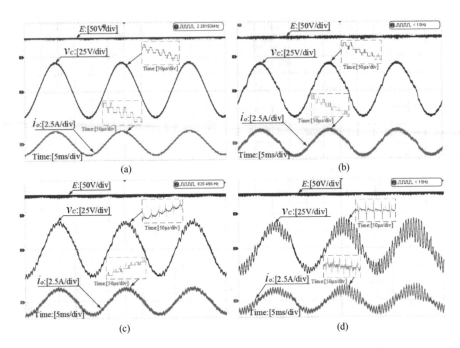

Fig. 4.30 Steady-state waveforms under resistive load condition for different K when $k_i = 20$ and $k_p = 0.04$. (**a**) $K = 0.542$; (**b**) $K = 0.642$; (**c**) $K = 0.742$; (**d**) $K = 0.842$

that the PWM inverter is nearly critically stable. And when K increases to 0.742, noticeable oscillations are observed in the waveforms, which indicates that the PWM inverter becomes unstable. And when $K = 0.842$, significant oscillations appear in waveforms of v_C and i_o, and that means the PWM inverter is highly unstable in this case.

Thus, the experimental results in Fig. 4.30 show consistency with the theoretical results, that is, the PWM inverter becomes unstable when $K > 0.742$ for $k_i = 20$ and $k_p = 0.04$. In addition, the presented magnified waveforms of v_C and i_o also show similar dynamic properties like those in the normal waveforms, and verify the consistency of the fast- and slow-scale stabilities for the PWM inverter.

4.6.2   Experimental Results Under Inductive-Resistive Load Condition

The steady-state waveforms under inductive-resistive load condition for different k_p when $k_i = 20$ and $K = 0.5$ are shown in Fig. 4.31. It can be seen that waveforms of filter capacitor voltage v_C and output current i_o are sinusoidal and periodic when $k_p = 0.05$, which indicates that the PWM inverter is stable. But, when $k_p = 0.06$, waveforms of v_C and i_o become slightly distorted, which means that the PWM inverter is almost critically stable. And when $k_p = 0.07$, waveforms of v_C and i_o

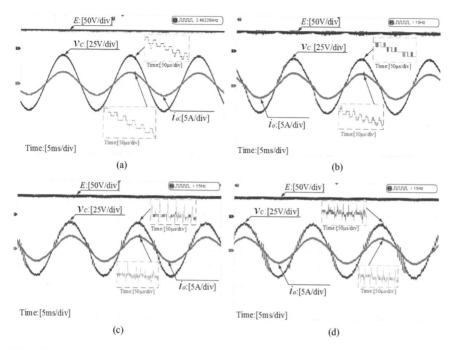

Fig. 4.31 Steady-state waveforms under inductive-resistive load condition for different k_p when $k_i = 20$, $K = 0.5$. (**a**) $k_p = 0.05$; (**b**) $k_p = 0.06$; (**c**) $k_p = 0.07$; (**d**) $k_p = 0.08$

become oscillating, which means that the PWM inverter becomes unstable. When $k_p = 0.08$, waveforms of v_C and i_o oscillate remarkably, which indicates that the PWM inverter is totally unstable.

Thus, the experimental results in Fig. 4.31 show good conformity with the theoretical result that the PWM inverter becomes unstable when $k_p > 0.07$ for $k_i = 20$ and $K = 0.5$. Furthermore, as can be observed in Fig. 4.31, the magnified and normal waveforms of v_C and i_o are also substantially consistent in dynamic characteristics, which proves the concordance of the fast- and slow-scale stabilities for the inverter under inductive-resistive load condition.

Figure 4.32 depicts the transient waveforms under inductive-resistive load condition when $k_i = 20$, $K = 0.5$, and $k_p = 0.04$, including both the transient waveforms in response to no load to nominal resistive load step change, and transient waveforms in response to -50% step change of load resistor. It can be seen that the dynamic performance of the PWM inverter with inductive-resistive load is also excellent.

4.6.3 Experimental Results Under Nonlinear Load Condition

Figure 4.33 presents the steady-state waveforms under nonlinear load condition with and without using harmonic control scheme for the 3rd, 5th, 7th, 9th, and 11th harmonic components. As shown in Fig. 4.33b, by employing the harmonic control

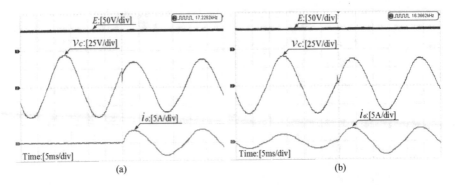

Fig. 4.32 Transient waveforms under inductive-resistive load condition when $k_i = 20$, $K = 0.5$ and $k_p = 0.04$ (**a**) Transient waveforms in response to no load to nominal resistive load step change; (**b**) Transient waveforms in response to -50% step change of load resistor

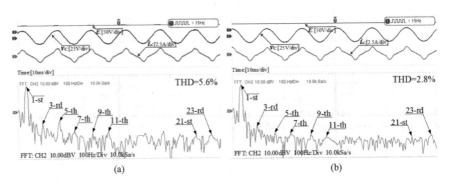

Fig. 4.33 Steady-state waveforms under nonlinear load condition with and without using harmonic control scheme. (a) Without using harmonic control scheme. (b) With using harmonic control scheme

scheme, the 3rd, 5th, 7th, 9th, and 11th harmonic components are significantly reduced. The total harmonic distribution (THD) of v_C becomes smaller, and approximately sinusoidal waveform of v_C is obtained, which validates the effectiveness of the proposed harmonic control scheme.

The steady-state waveforms under nonlinear load condition for different k_p when $k_i = 20$ and $K = 0.5$ are shown in Fig. 4.34. In Fig. 4.34a, when $k_p = 0.038$ which is lower than the critical value 0.048 in Fig. 4.11, the THD of v_C is relatively small. But, when k_p increases to 0.058, the THD of v_C becomes much higher, as shown in Fig. 4.34b. The experimental results reveal that the Lyapunov exponent method and equivalent model of diode rectifier are effective for the approximate slow-scale stability analysis under nonlinear load condition.

Figure 4.35 shows the steady-state waveforms under nonlinear load condition for different K when $k_i = 20$ and $k_p = 0.04$. It can be seen that, when $K = 0.452$ which is lower than the critical value 0.552 in Fig. 4.15, the harmonic distortion of v_C is less

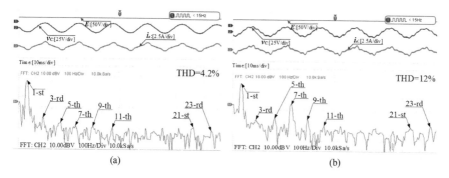

Fig. 4.34 Steady-state waveforms under nonlinear load condition for different k_p when $k_i = 20$ and $K = 0.5$ for different k_p. (**a**) $k_p = 0.038$. (**b**) $k_p = 0.058$

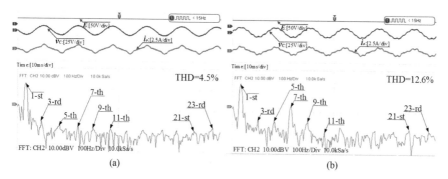

Fig. 4.35 Steady-state waveforms under nonlinear load condition for different K when $k_i = 20$ and $k_p = 0.04$ for different K. (**a**) $K = 0.452$. (**b**) $K = 0.652$

obvious. However, when $K = 0.652$, which is higher than 0.552, the harmonic distortion of v_C increases significantly, as shown in Fig. 4.35b. The experimental results also clearly verify the validity of the Lyapunov exponent method and equivalent model of diode rectifier for the approximate slow-scale stability analysis under nonlinear load condition.

4.7 Conclusion

This chapter presents the stability analysis of a digital controlled single-phase VSI with SRF voltage control by employing two nonlinear approaches, the Jacobian matrix method and the Lyapunov exponent method. To adopt these two analysis methods, the stroboscopic model of the single-phase PWM inverter is established by using the state-space averaging technique. The analyses are subsequently implemented under variations of three control parameters of voltage loop and current loop, and stability regions of the PWM inverter system are obtained.

In addition, for the derived stroboscopic model, the Jacobian matrix method and the Lyapunov exponent method are proved to be mathematically equivalent. Therefore, the fast-scale stability and slow-scale stability described by the Jacobian matrix method and the Lyapunov exponent method, respectively, are consistent for the studied PWM inverter in stand-alone mode. Moreover, the simulation results obtained from EMTP software are presented to study the effect of control delay, load parameters, and controller gain on stability characteristics of the closed-loop system. The theoretical results are verified by the experimental results from reduced-scale prototype system, which indicates that discrete-time model plus the Jacobian matrix method or the Lyapunov exponent method are capable to analyze the stability of a switching converter with SRF control loops accurately. The presented approach can also be applied to other converter topologies either in the standalone or grid-connected operating mode.

Appendix A: Expressions of $\alpha, \beta, K_1, K_2, K_3, K_4$ in Eq. (4.5)

$$\alpha = -\frac{1}{2RC} \tag{4.A1}$$

$$\beta = \sqrt{\frac{1}{LC} - \frac{1}{(2RC)^2}} \tag{4.A2}$$

$$K_1 = i_L(n) + \frac{1}{R}\left[1 - 2d(n)\right]E \tag{4.A3}$$

$$K_2 = -\frac{v_C(n) + \left[1 - 2d(n)\right]E}{\beta L} - \frac{\alpha}{\beta R}\left\{Ri_L(n) + \left[1 - 2d(n)\right]E\right\} \tag{4.A4}$$

$$K_3 = v_C(n) + \left[1 - 2d(n)\right]E \tag{4.A5}$$

$$K_4 = \left(\frac{\alpha^2 L}{\beta} + \beta L\right)i_L(n) + \frac{\alpha}{\beta}v_C(n) + \frac{\alpha R + \alpha^2 L + \beta^2 L}{\beta R}\left[1 - 2d(n)\right]E \tag{4.A6}$$

Appendix B: Definitions of Coefficients in Eq. (4.8)

$$r = \sqrt[3]{-\frac{q}{2} + \sqrt{\frac{q^2}{4} + \frac{p^3}{27}}} + \sqrt[3]{-\frac{q}{2} - \sqrt{\frac{q^2}{4} + \frac{p^3}{27}}} - \frac{R_1}{3L_1} \tag{4.B1}$$

$$\alpha_1 = -\frac{1}{2}\left(\sqrt[3]{-\frac{q}{2} + \sqrt{\frac{q^2}{4} + \frac{p^3}{27}}} + \sqrt[3]{-\frac{q}{2} - \sqrt{\frac{q^2}{4} + \frac{p^3}{27}}} \right) - \frac{R_1}{3L_1} \tag{4.B2}$$

$$\beta_1 = \frac{\sqrt{3}}{2}\left(\sqrt[3]{-\frac{q}{2} + \sqrt{\frac{q^2}{4} + \frac{p^3}{27}}} - \sqrt[3]{-\frac{q}{2} - \sqrt{\frac{q^2}{4} + \frac{p^3}{27}}} \right) \tag{4.B3}$$

$$p = \frac{L + L_1}{LL_1 C} - \frac{R_1^2}{3L_1^2} \tag{4.B4}$$

$$q = \frac{R_1}{LL_1 C} - \frac{2R_1^3}{27L_1^3} - \frac{R_1(L + L_1)}{3LL_1^2 C} \tag{4.B5}$$

$$K_5 = \left[-\frac{1}{\theta LC} + \frac{\beta_1^2 - \alpha_1^2}{\theta} - \frac{\alpha_1 L}{\theta} \right] i_L(n) - \frac{2\alpha_1}{\theta L} v_C(n) + \frac{1}{\theta LC} i_o(n) - \frac{2\alpha_1 [2d(n)-1]}{\theta L} \frac{E}{B} \tag{B}$$

$$K_6 = i_L(n) - K_5 \tag{4.B7}$$

$$K_7 = \frac{1}{L\beta_1}\left\{ [2d(n)-1]E - v_C(n) - \alpha_1 Li_L(n) \right\} - \frac{r - \alpha_1}{\beta_1} K_5 \tag{4.B8}$$

$$\theta = r^2 - 2\alpha_1 + \alpha_1^2 - \beta_1^2 \tag{4.B9}$$

Appendix C: Expressions of Matrix Elements in Eq. (4.13)

$$\frac{\partial i_L(n+1)}{\partial i_L(n)} = e^{\alpha T}\left(\cos \beta T - \frac{\alpha}{\beta}\sin \beta T \right) \tag{4.C1}$$

$$\frac{\partial i_L(n+1)}{\partial v_C(n)} = -\frac{1}{\beta L} e^{\alpha T} \sin \beta T \tag{4.C2}$$

$$\frac{\partial i_L(n+1)}{\partial d(n)} = \frac{2E}{R} \left[e^{\alpha T} \left(\frac{\alpha L + R}{\beta L} \sin \beta T - \cos \beta T \right) + 1 \right] \tag{4.C3}$$

$$\frac{\partial v_C(n+1)}{\partial i_L(n)} = \frac{\alpha^2 L + \beta^2 L}{\beta} e^{\alpha T} \sin \beta T \tag{4.C4}$$

$$\frac{\partial v_C(n+1)}{\partial v_C(n)} = e^{\alpha T} \left(\cos \beta T + \frac{\alpha}{\beta} \sin \beta T \right) \tag{4.C5}$$

$$\frac{\partial v_C(n+1)}{\partial d(n)} = 2E \left[1 - e^{\alpha T} \left(\cos \beta T + \frac{\alpha R + \alpha^2 L + \beta^2 L}{\beta R} \sin \beta T \right) \right] \tag{4.C6}$$

$$\frac{\partial d(n+1)}{\partial i_L(n)} = -\frac{1}{2} K \tag{4.C7}$$

$$\frac{\partial d(n+1)}{\partial v_C(n)} = \frac{1}{2} K \left(\frac{1}{R} - k_p - k_i T \right) \tag{4.C8}$$

$$\frac{\partial d(n+1)}{\partial d(n)} = 0 \tag{4.C9}$$

References

1. Wai, R. J., Lin, C. Y., Huang, Y. C., & Chang, Y. R. (2013). Design of high-performance standalone and grid-connected inverter for distributed generation applications. *IEEE Transactions on Industrial Electronics, 60*(4), 1542–1555.
2. Trujillo, C., Velasco, D., Garcera, G., Figueres, E., & Guacaneme, J. (2013). Reconfigurable control scheme for a PV microinverter working in both grid-connected and island modes. *IEEE Transactions on Industrial Electronics, 60*(4), 1582–1595.
3. Deng, H., Oruganti, R., & Srinivasan, D. (2008). A simple control method for high-performance UPS inverters through output-impedance reduction. *IEEE Transactions on Industrial Electronics, 55*(02), 888–898.
4. Zhang, X., & Spencer, J. W. (2012). Study of multisampled multilevel inverters to improve control performance. *IEEE Transactions on Power Electronics, 27*(11), 4409–4416.
5. Komurcugil, H., Altin, N., Ozdemir, S., & Sefa, I. (2015). An extended Lyapunov-function-based control strategy for single-phase UPS inverters. *IEEE Transactions on Power Electronics, 30*(7), 3976–3983.

6. Meza, C., Biel, D., Jeltsema, D., & Scherpen, J. M. A. (2012). Lyapunov based control scheme for single phase grid-connected PV central inverters. *IEEE Transactions on Control Systems Technology, 20*(2), 520–529.
7. Komurcugil, H., Altin, N., Ozdemir, S., & Sefa, I. (2016). Lyapunov-function and proportional-resonant based control strategy for single-phase grid-connected VSI with LCL filter. *IEEE Transactions on Industrial Electronics, 63*(5), 2838–2849.
8. Rezkallah, M., Sharma, S., Chandra, A., Singh, B., & Rousse, D. R. (2017). Lyapunov function and sliding mode control approach for solar-PV grid Interface system. *IEEE Transactions on Industrial Electronics, 64*(1), 785–795.
9. Wolf, A., Swift, J. B., Swinney, H. L., & Vastano, J. A. (1985). Determining Lyapunov exponents from a time series. *Physica D., 16*(3), 285–317.
10. Saublet, L.-M., Gavagsaz-Ghoachani, R., Martin, J.-P., Nahid-Mobarakeh, B., & Pierfederici, S. (2016). Asymptotic stability analysis of the limit cycle of a cascaded DC–DC converter using sampled discrete-time modeling. *IEEE Transactions on Industrial Electronics, 63*(4), 2477–2487.
11. Khalil, H. K. (1996). *Nonlinear systems* (2nd ed.). Prentice-Hall.
12. Tse, C. K. (2004). *Complex behavior of switching power converters* (1st ed.). CRC Press.
13. Zhang, X., Xu, J., Bao, B., & Zhou, G. (2016). Asynchronous-switching map-based stability effects of circuit parameters in fixed off-time controlled buck converter. *IEEE Transactions on Power Electronics, 31*(9), 6686–6697.
14. di Bernardo, M., & Vasca, F. (2000). Discrete-time maps for the analysis of bifurcations and Chaos in DC/DC converters. *IEEE Transactions on Power Electronics, 47*(02), 130–143.
15. Chen, M., Zheng, C., Qian, Z., & Yuan, X. (2005). The analysis of inverter under rectifier load using a nonlinear rectifier model. *Proceedings of INTELEC, 28*(15), 455–459.
16. Han, Y., Fang, X., Yang, P., Xu, L., & Guerrero, J. M. (2018). Stability analysis of digital-controlled single-phase inverter with synchronous reference frame voltage control. *IEEE Transactions on Power Electronics, 33*(7), 6333–6350.

Chapter 5
Small-Signal Modeling and Controller Synthesis of BPF-Based Droop Control for Single-Phase Islanded Microgrid

In the hierarchical control framework of microgrid, by introducing a feedback path in the conventional droop control loop of islanded microgrid, the voltage amplitude and frequency restoration can be effectively achieved by using the secondary control loop. However, the use of low bandwidth communication (LBC) links, the communication delays and the underlying data drop, would increase the system complexity and reduce its reliability. In our previous studies, it is found that the voltage and frequency deviation of islanded three-phase microgrid in synchronous reference frame can be restored by utilizing a band-pass filter (BPF)-based improved droop control strategy, without any communication lines and additional control loops. For more general scenarios, the BPF-based droop control scheme is first extended to islanded single-phase microgrid in hybrid frame in this chapter to achieve voltage amplitude and frequency deviation restoration. Moreover, the dynamic stability of the studied system is addressed by the derived reduced-order small-signal model, which simplifies the modeling process and theoretical analysis. Followed by the system model, the impact of system parameters variation on the stability and dynamic performance of the microgrid are subsequently predicted by applying the eigenvalue-based analysis approach. Consequently, the analytical results show an overdamped system with good stability and robustness against system parameters drift. Finally, the effectiveness of eigenvalue analysis is verified by extensive simulation results obtained from PLECS and EMTP, and the experiment results are provided to further validate the effectiveness of the BPF-based droop control method in islanded single-phase microgrid.

5.1 Introduction

In an islanded microgrid, the droop control is capable to realize power sharing among parallel DG units by imitating the external characteristics of synchronous generators. The conventional droop control based on local measurement is simple

© Springer Nature Switzerland AG 2022
Y. Han, *Modeling and Control of Power Electronic Converters for Microgrid Applications*, https://doi.org/10.1007/978-3-030-74513-4_5

implementation and low cost due to the absence of communication links. However, the droop control is subjected to the steady-state error when the load is abruptly changed, since the output voltage amplitude and frequency of inverter deviate from their rated values [1–3]. Accordingly, a supplemented secondary control strategy can be adopted to restore the voltage amplitude and frequency deviation caused by conventional droop control. In [4], with the hypothesis of a low-voltage microgrid, a consensus-based distributed secondary control is introduced to improve the accuracy of voltage regulation. In [5], a distributed model predictive secondary voltage control is utilized to reduce the convergence time of voltage amplitude. Besides, a distributed secondary voltage control based on RBF-neural-network sliding mode technique is designed [6] to diminish the voltage fluctuation in the process of voltage regulation. Further, a distributed noise-resilient secondary voltage and frequency control scheme is proposed in [7] for deviation restoration, with the consideration of additional noise disturbance in measurement. Although the robustness and stability of system are enhanced by means of aforementioned improved secondary control, the reliability of microgrid is still affected by the usage of communication lines. Moreover, the slightly different time signals among inverters resulting from clock drifts, the uncertain communication delays, data drop or communication failure may also reduce microgrid reliability.

Referring to [8], a washout filter-based droop control scheme is able to restore voltage amplitude and frequency deviation, without any communication lines and additional control loops. Further, a washout filter-based power sharing method is combined with a model predictive control (MPC) in [9], where the MPC is employed to stabilize the dc-bus voltage for improving power quality. Actually, the equivalence between the ideal secondary control and the washout filter-based control has been manifested in [10], and the washout filter-based droop control can be regarded as a BPF-based improved droop control. Whereas the previous research in [10] is based on three-phase systems in dq-frame, and the feasibility of BPF-based droop control has not been explored in single-phase microgrid controlled in hybrid frame. Furthermore, the full-order model of complete microgrid is time-consuming for modeling process, and the dynamic stability of microgrid using BPF-based droop control has not been adequately explored.

In order to devise a generic BPF-based droop control scheme both in three-phase and single-phase microgrid, the application of BPF-based droop control in single-phase microgrid is discussed in this chapter, and the stability and dynamic performance analysis in this scenario is also fully investigated. As for modeling the inverter-based microgrid, since the microgrid is a complex system including multiple timescale control interactions, thus the detailed full-order model is computationally expensive due to its high-order characteristics. Therefore, it is necessary to perform reasonable model order reduction to facilitate subsequent theoretical analysis, and the relevant studies have been published for solving this issue. In [11–13], the fast and slow timescale state variables are separated by using singular perturbation method, which implements model order reduction by neglecting the fast state variables. In [14–16], the model order of microgrid is reduced using aggregate equivalent model with the assumption of similar dynamical inverters. In [17, 18], the Kron reduction is utilized to simplify system model. However, the full-order model of system is originally required before executing the aforementioned model order reduction, which increases the complexity of modeling process. In the previous literatures, it has been found that the fast states of inverter are mainly related to inner control loops and inverter output filter, and the slow states are correlated with outer droop control loop and dc-link capacitors.

Notably, the preliminary results of this chapter were presented in IEEE SPEC 2018 conference [19]. However, only the active droop control coefficient, the cutoff frequency of high-pass filter (HPF) and the load resistance are considered, this chapter aims to extend the analysis to multiple parameters variation scenarios by using eigenvalue-based approach. The BPF-based droop control approach is extended to single-phase inverters with synchronous reference frame (SRF) voltage control, and the feasibility and effectiveness of this method in islanded single-phase microgrid are confirmed by experimental results. Compared with conventional droop control, the voltage amplitude and frequency can be effectively restored to reference values in case of load changes, without any communication lines and additional control loops. Moreover, the reduced-order small-signal model of stand-alone single-phase microgrid with BPF-based droop control is derived, and the impact of system parameters shifting, including the active and reactive droop control coefficients, the bandwidth of BPF and the load resistance, on the stability and dynamic performance of system are analytically predicted by using eigenvalue loci of the system matrix, which provide a basis for the parameter selection [20, 21].

The rest of this chapter is organized as follows. Section 5.2 derives the reduced-order small-signal model of islanded single-phase microgrid with BPF-based droop control scheme. Section 5.3 predicts the stability and dynamic performances of microgrid system by using the established system matrix model. Besides, Sects. 5.4 and 5.5 present the simulation results from PLECS and EMTP software to confirm the effectiveness of eigenvalue analysis. And Sect. 5.6 provides the experimental results from a small-scale prototype system to verify the effectiveness of the BPF-based droop control in a single-phase microgrid, and Sect. 5.7 concludes this chapter.

5.2 Small-Signal Modeling of Single-Phase Islanded Microgrid

The block diagram of the studied islanded single-phase microgrid in hybrid frame with the BPF-based droop control method is shown in Fig. 5.1. The main circuit includes the inverters, LC filters, line impedances and a resistive load connected as a common bus. Note that the output power of inverter is filtered by a BPF in this chapter instead of a low-pass filter (LPF). Moreover, a proportional integral (PI) controller is adopted for synchronous reference frame (SRF) voltage control, which realizes the zero steady-state error tracking of AC signals.

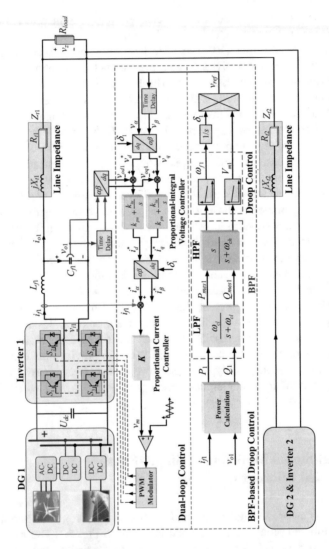

Fig. 5.1 Studied islanded microgrid composed of two parallel single-phase inverters

In order to implement the stability analysis of islanded single-phase microgrid system, a reduced-order small-signal model of the system is derived in this section, which consists of two parts, including the active and reactive power of inverters and the BPF-based droop control loop.

5.2.1   Active and Reactive Power Controller

When the islanded microgrid composed of two parallel-connected single-phase inverters operating in the steady-state, the inverters can be equivalent to ideal voltage sources seen from the filter output terminal. Each inverter can be described by its line current and terminal states, such as voltage amplitude, phase, and angular frequency, neglecting the internal states of inverter, and thus the equivalent circuit of islanded single-phase microgrid can be denoted by Fig. 5.2.

In Fig. 5.2, the V_{m1} and V_{m2} are the amplitudes of inverter output voltage v_{o1} and v_{o2}, respectively, the δ_1 and δ_2 are the phase deviations between the inverter output voltage and the load voltage, and V_{mz} is the amplitude of load voltage. Moreover, the Z_{t1} and Z_{t2} are the line impedance between the filter output terminal and the common load, and Z_{load} and i_{load} are the load impedance and load current, respectively. In order to simplify the theoretical analysis, it is assumed that $Z_{t1} = Z_{t2} = Z_t = R_t + jX_t$, that is, $\theta_t = \arctan(X_t/R_t)$.

Generally, the amplitude and frequency of the inverter output voltage are utilized to implement droop control. Hence, the phasors of voltage and current are represented by their own amplitude and phase to facilitate the following small signal modeling and theoretical analysis. In this way, the phasor of output current i_{o1} of inverter 1 around the steady-state operating point (V_{m1e}, V_{m2e}, δ_{1e}, δ_{2e}, ω_{f1e}, ω_{f2e}) can be defined as [19, 20]

$$i_{ole} = \frac{V_{m1e}\angle\delta_{1e} - V_{mze}\angle 0}{|Z_t|\angle\theta_t} = \frac{V_{m1e}}{|Z_t|}\angle(\delta_{1e} - \theta_t) - \frac{V_{mze}}{|Z_t|}\angle -\theta_t \tag{5.1}$$

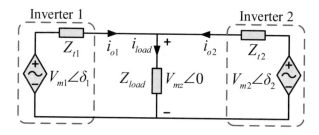

Fig. 5.2 Equivalent circuit of the islanded single-phase microgrid composed of two parallel-connected inverters

where $|Z_t|$ and θ_t are the magnitude and phase angle of the line impedance, respectively, and the subscripts "e" represents the steady-state values of variables.

Similarly, the phasor of output current i_{o2} of inverter 2 in the steady-state can be described as

$$i_{o2e} = \frac{V_{m2e}\angle\delta_{2e} - V_{mze}\angle 0}{|Z_t|\angle\theta_t} = \frac{V_{m2e}}{|Z_t|}\angle(\delta_{2e} - \theta_t) - \frac{V_{mze}}{|Z_t|}\angle - \theta_t \quad (5.2)$$

Meanwhile, as shown in Fig. 5.2, eq. (5.3) can be obtained according to Kirchhoff's voltage law (KVL) and voltage–current relationship.

$$\left(i_{o1e} + i_{o2e}\right)\cdot\left(|Z_{load}|\angle\theta_{load}\right) = V_{mze}\angle 0 \quad (5.3)$$

where Z_{load} is selected as a purely resistive load, $Z_{load} = R_{load}$, that is, $\theta_{load} = 0$.

Substituting (5.1) and (5.2) into (5.3), the magnitude of load voltage can be given as

$$V_{mze} = \frac{V_{m1e}\cos(\delta_{1e} + \theta_{load} - \theta_t)}{2\cos(\theta_{load} - \theta_t) + (|Z_t|/|Z_{load}|)} + \frac{V_{m2e}\cos(\delta_{2e} + \theta_{load} - \theta_t)}{2\cos(\theta_{load} - \theta_t) + (|Z_t|/|Z_{load}|)} \quad (5.4)$$

Hence, by combining (5.1)–(5.4), the output active and reactive power of inverter 1 and inverter 2 can be calculated as

$$P_{1e} = \frac{V_{m1e}^2}{2|Z_t|}\cos\theta_t - \frac{V_{m1e}V_{mze}}{2|Z_t|}\cos(\delta_{1e} + \theta_t) \quad (5.5)$$

$$P_{2e} = \frac{V_{m2e}^2}{2|Z_t|}\cos\theta_t - \frac{V_{m2e}V_{mze}}{2|Z_t|}\cos(\delta_{2e} + \theta_t) \quad (5.6)$$

$$Q_{1e} = \frac{V_{m1e}^2}{2|Z_t|}\sin\theta_t - \frac{V_{m1e}V_{mze}}{2|Z_t|}\sin(\delta_{1e} + \theta_t) \quad (5.7)$$

$$Q_{2e} = \frac{V_{m2e}^2}{2|Z_t|}\sin\theta_t - \frac{V_{m2e}V_{mze}}{2|Z_t|}\sin(\delta_{2e} + \theta_t) \quad (5.8)$$

Due to uncertain small-signal disturbances around the steady-state operating point, the linearized models of (5.5)–(5.8) can be obtained as

$$\Delta P_{1e} = \frac{\partial P_{1e}}{\partial V_{m1e}}\Delta V_{m1e} + \frac{\partial P_{1e}}{\partial \delta_{1e}}\Delta\delta_{1e} + \frac{\partial P_{1e}}{\partial V_{m2e}}\Delta V_{m2e} + \frac{\partial P_{1e}}{\partial \delta_{2e}}\Delta\delta_{2e} \quad (5.9)$$

$$\Delta P_{2e} = \frac{\partial P_{2e}}{\partial V_{m1e}} \Delta V_{m1e} + \frac{\partial P_{2e}}{\partial \delta_{1e}} \Delta \delta_{1e} + \frac{\partial P_{2e}}{\partial V_{m2e}} \Delta V_{m2e} + \frac{\partial P_{2e}}{\partial \delta_{2e}} \Delta \delta_{2e} \qquad (5.10)$$

$$\Delta Q_{1e} = \frac{\partial Q_{1e}}{\partial V_{m1e}} \Delta V_{m1e} + \frac{\partial Q_{1e}}{\partial \delta_{1e}} \Delta \delta_{1e} + \frac{\partial Q_{1e}}{\partial V_{m2e}} \Delta V_{m2e} + \frac{\partial Q_{1e}}{\partial \delta_{2e}} \Delta \delta_{2e} \qquad (5.11)$$

$$\Delta Q_{2e} = \frac{\partial Q_{2e}}{\partial V_{m1e}} \Delta V_{m1e} + \frac{\partial Q_{2e}}{\partial \delta_{1e}} \Delta \delta_{1e} + \frac{\partial Q_{2e}}{\partial V_{m2e}} \Delta V_{m2e} + \frac{\partial Q_{2e}}{\partial \delta_{2e}} \Delta \delta_{2e} \qquad (5.12)$$

where the symbol Δ represents the small-signal deviation of the variables relative to the steady-state operating point.

5.2.2   BPF-Based Enhanced Droop Controller

It is noteworthy that the dynamic performance of system is dominated by the outer droop control loop due to its much lower control bandwidth than that of inner voltage and current control loops, that is, assuming no mutual coupling between outer and inner loops; thus, the impact of dual-loop controllers on system dynamics can be neglected. In this case, the inverter can be regarded as an ideal voltage source, that is, the reference voltage synthesized by BPF-based droop control is equal to the actual output voltage of inverter. Specifically, the linearized model of power loop with BPF-based droop control is shown in Fig. 5.3.

Commonly, the low-pass filter (LPF) is employed to measure the output power of inverter. Moreover, the high-pass filter (HPF) can suppresses the DC component and passes the transient component of signals, and it also makes the system more robust at the steady-state operating points. It is assumed that the cutoff frequencies of the LPF of two inverters are consistent with each other and, therefore, the linearized model of active and reactive output power of inverters measured by LPF in s-domain can be denoted as

$$\Delta P_{1mese}(s) = \frac{\omega_{cl}}{s + \omega_{cl}} \cdot \Delta P_{1e}(s) \qquad (5.13)$$

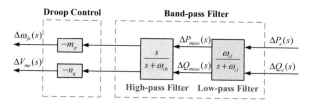

Fig. 5.3 The linearized model of power loop with BPF-based droop control

$$\Delta P_{2mese}(s) = \frac{\omega_{cl}}{s + \omega_{cl}} \cdot \Delta P_{2e}(s) \tag{5.14}$$

$$\Delta Q_{1mese}(s) = \frac{\omega_{cl}}{s + \omega_{cl}} \cdot \Delta Q_{1e}(s) \tag{5.15}$$

$$\Delta Q_{2mese}(s) = \frac{\omega_{cl}}{s + \omega_{cl}} \cdot \Delta Q_{2e}(s) \tag{5.16}$$

where ω_{cl} is the cutoff angular frequency of LPF.

Transforming (5.13)–(5.16) into the time-domain, these equations can be rewritten as

$$\Delta \dot{P}_{1mese} = \omega_{cl}\left(\Delta P_{1e} - \Delta P_{1mese}\right) \tag{5.17}$$

$$\Delta \dot{P}_{2mese} = \omega_{cl}\left(\Delta P_{2e} - \Delta P_{2mese}\right) \tag{5.18}$$

$$\Delta \dot{Q}_{1mese} = \omega_{cl}\left(\Delta Q_{1e} - \Delta Q_{1mese}\right) \tag{5.19}$$

$$\Delta \dot{Q}_{2mese} = \omega_{cl}\left(\Delta Q_{2e} - \Delta Q_{2mese}\right) \tag{5.20}$$

According to Fig. 5.3, the dynamic models of voltage and frequency controllers can be obtained as

$$\Delta \omega_{fe}(s) = -m_p \Delta P_{mese}(s) \cdot \frac{s}{s + \omega_{ch}} \tag{5.21}$$

$$\Delta V_{me}(s) = -n_p \Delta Q_{mese}(s) \cdot \frac{s}{s + \omega_{ch}} \tag{5.22}$$

where m_p and n_q are the droop control coefficients of active and reactive power, respectively.

With respect to inverter 1 and inverter 2, we can transform (5.21) and (5.22) into time-domain, which can be modified as

$$\Delta \dot{\omega}_{f1e} = -m_p \Delta \dot{P}_{1mese} - \omega_{ch} \Delta \omega_{f1e} \tag{5.23}$$

$$\Delta \dot{\omega}_{f2e} = -m_p \Delta \dot{P}_{2mese} - \omega_{ch} \Delta \omega_{f2e} \tag{5.24}$$

$$\Delta \dot{V}_{m1e} = -n_q \Delta \dot{Q}_{1mese} - \omega_{ch} \Delta V_{m1e} \tag{5.25}$$

$$\Delta \dot{V}_{m2e} = -n_q \Delta \dot{Q}_{2mese} - \omega_{ch} \Delta V_{m2e} \tag{5.26}$$

where ω_{ch} is the cutoff angular frequency of HPF.

In addition, the angular frequencies of inverter 1 and inverter 2 are expressed by the first-order derivative of phase as

$$\Delta\dot{\delta}_1 = \Delta\omega_{f1} \tag{5.27}$$

$$\Delta\dot{\delta}_2 = \Delta\omega_{f2} \tag{5.28}$$

5.2.3   Complete Small-Signal Model of Single-Phase Islanded Microgrid

By combining (5.17)–(5.20) and (5.23)–(5.28), the small-signal model of the studied islanded single-phase microgrid with BPF-based droop control method can be derived as [19, 20]

$$\Delta\dot{\mathbf{X}} = \mathbf{A}\Delta\mathbf{X} \tag{5.29}$$

where $\Delta\mathbf{X}$ and $\mathbf{A}$ are the state vector and system matrix, respectively, which are shown in Appendix.

Based on the established small-signal model in (5.29), the stability and dynamic performance of microgrid are analyzed by using eigenvalue analysis method in Sect. 5.3.

5.3   Stability and Dynamic Performance Analysis

In this section, the stability and dynamic performance of microgrid are predicted by plotting the eigenvalue variation loci of system matrix. The system parameters and equilibrium point are listed in Table 5.1.

Taking the active and reactive droop control coefficients, the bandwidth of BPF and the load resistance as variables, the locus of the eigenvalue λ_i ($i = 1, 2, 3, ..., 9, 10$) can be plotted to predict the dynamic performance and stability of system. It is noteworthy that the system stability is mainly related to the eigenvalues near imaginary axis, and the damping ratio of system is dominated by the eigenvalues closed to real axis.

Table 5.1 System parameters and equilibrium point

System parameters		Values
R_t		2 Ω
X_t		0.628 Ω
R_{load}		10 Ω
ω_{cl}		5 × 2π rad/s
ω_{ch}		2 × 2π rad/s
m_p		0.01 rad/s/W
n_q		0.01 V/Var
L_{f1}, L_{f2}		2 mH
C_{f1}, C_{f2}		2.2 μF
DC-link voltage	DG1	39.8 V
	DG2	40 V
Voltage controller	Proportional coefficient (k_{pu})	1
	Integral coefficient (k_{iu})	10
Current controller	Proportional coefficient (k_{pc})	0.3
Equilibrium point		Values
V_{m1}, V_{m2}		20 V
δ_1, δ_2		0.031416 rad
ω_{f1}, ω_{f2}		314 rad/s

5.3.1   Effect of Active Power Droop Coefficient

In this subsection, the eigenvalues sensitivity to active droop control coefficient m_p is investigated. When m_p is changed from 0.0001 to 0.05, three typical values of f_{ch} are selected, that is, 1, 3, 4 Hz, and the rest of system parameters remain unchanged. The eigenvalue loci of system matrix are shown in Fig. 5.4.

As shown in Fig. 5.4, no eigenvalues are located in right half plane, which manifests that a stable system is obtained in the variation range of m_p. However, when f_{ch} is a certain value, the stability margin of system is decreased as m_p increases. In addition, from Fig. 5.4a–c, λ_5, λ_6, λ_7, and λ_8 travel away from the real axis when enlarging f_{ch}, which indicates the damping ratios of corresponding response components are continuously reduced. Note that the damping ratio of system is dominated by two pairs of unequal negative real roots (λ_3 and λ_{10}, λ_4 and λ_9) moved along the real axis, which is decreased with the improved dynamic performance of system.

5.3.2   Effect of Reactive Power Droop Coefficient

In this part, the eigenvalue loci are shown in Fig. 5.5 when the reactive droop control coefficient n_q is varied from 0.0001 to 0.05. Simultaneously, three different active droop control coefficients m_p are taken into account. It can be seen that the system is stable since the all eigenvalues are in the left half plane. However, in case

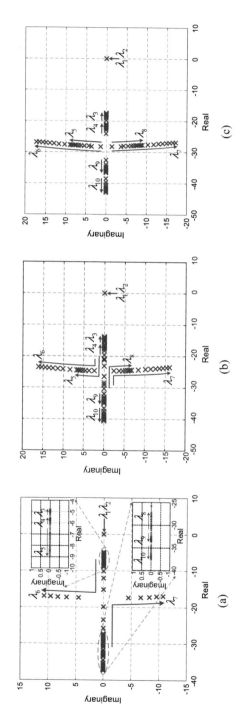

Fig. 5.4 Eigenvalue loci of the system matrix with the variation of active droop control coefficient m_p (0.0001 $\leq m_p \leq$ 0.05). (**a**) f_{ch} =1 Hz. (**b**) f_{ch} =3 Hz. (**c**) f_{ch} =4 Hz

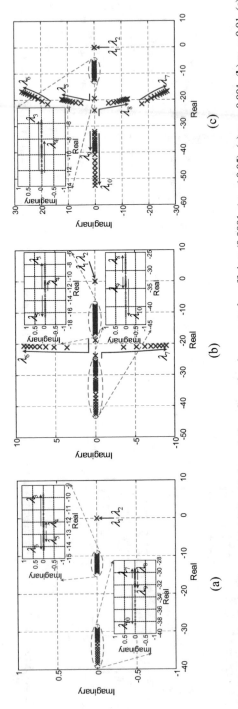

Fig. 5.5 Eigenvalue loci of the system matrix with the variation of reactive droop control coefficient ($0.0001 \leq n_q \leq 0.05$). (**a**) $m_p = 0.001$. (**b**) $m_p = 0.01$. (**c**) $m_p = 0.05$

of a certain m_p, the stability margin of system is gradually reduced as the n_q enlarges. Besides, from Fig. 5.5a–c, λ_5, λ_6, λ_7 and λ_8 move away from the real axis when enlarging m_p, which represents the partial response components of system. Similarly, the dynamic response of overall system is mainly related to the eigenvalues closed to real axis.

Consequently, the system is overdamped and the damping ratio is decreased according to the variation loci of two pairs of eigenvalues (λ_3 and λ_{10}, λ_4 and λ_9).

5.3.3 Effect of BPF Parameters

Taking three different values of the cutoff frequency f_{cl} of LPF, simultaneously, when the cutoff frequency f_{ch} of HPF is varied from 1 to 3 Hz, the eigenvalue loci of the system matrix can be obtained as shown in Fig. 5.6.

Similarly, no eigenvalues lie in the right half plane in Fig. 5.6, in which case the system is also stable. From Fig. 5.6a–c, when f_{cl} is a certain value, λ_5, λ_6, λ_7, and λ_8 move away from the real axis as f_{ch} enlarges. However, if the value of f_{cl} of group changing is larger, the terminus of eigenvalues λ_5, λ_6, λ_7, and λ_8 are closer to the real axis. Finally, λ_5, λ_6, λ_7, and λ_8 move on the real axis when $\omega_{cl} = 7$ Hz. It is noteworthy that most of the eigenvalues travel away from the imaginary axis apart from λ_1 and λ_2, which shows that the attenuation of the corresponding exponential component is gradually accelerated, that is, the time-domain response of the system is increasingly fast.

5.3.4 Effect of Load Transient Response

This subsection studies the eigenvalues sensitivity to the load resistance R_{load}. When the load resistance is changed from 2 to 40 Ω, three different m_p, that is, 0.0001, 0.001, 0.05, are simultaneously chosen. Sequentially, the eigenvalue loci of system matrix are illustrated in Fig. 5.7.

In Fig. 5.7, the system is still stable owing to no right-half-plane eigenvalues. With the increase of m_p, λ_5, λ_6, λ_7, and λ_8 travel away from real axis, which demonstrates the dynamic performance of corresponding response components is enhanced. Specifically, two pairs of eigenvalues (λ_3 and λ_{10}, λ_4 and λ_9) on real axis dominate the dynamic performance of system, and the damping ratio of system is reduced as load resistance R_{load} increases, with an overdamped system. It can be concluded from the above eigenvalue analysis, the BPF-controlled microgrid system exhibits slow dynamic response performance due to its overdamped characteristics. In order to guarantee fast dynamic response, a relatively large value of m_p and n_q should be selected, a relatively large frequency bandwidth of BPF should be chosen.

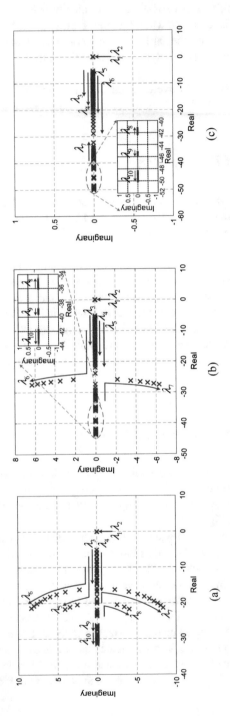

Fig. 5.6 Eigenvalue loci of the system matrix with the variation of the cutoff frequency of high-pass filter ($1\,\mathrm{Hz} \leq f_{ch} \leq 3\,\mathrm{Hz}$). (**a**) $f_{cl} = 4$ Hz. (**b**) $f_{cl} = 6$ Hz. (**c**) $f_{cl} = 7$ Hz

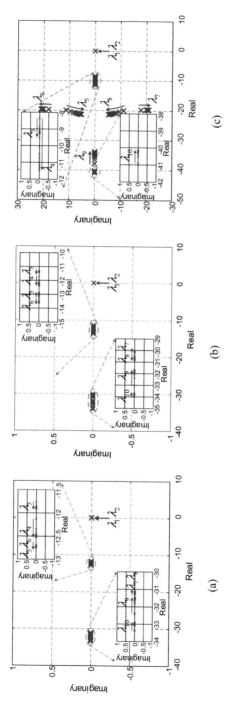

Fig. 5.7 Eigenvalue loci of the system matrix with the variation of load resistance ($2 \, \Omega \leq R_{load} \leq 40 \, \Omega$). (**a**) $m_p = 0.0001$. (**b**) $m_p = 0.001$. (**c**) $m_p = 0.05$

5.4   Simulation Results from PLECS

In this section, the dynamic performance of the microgrid system is verified by simulation results obtained from PLECS software. The adopted system parameters are listed in Table 5.1. By incorporating an extra load resistance (Load 2) at $t = 2$ s, the dynamic performance of the microgrid system with BPF-based droop control is investigated in this subsection.

Figure 5.8 shows the simulation results in case of different active droop control coefficients. Before $t = 2$ s, the load 1 of 10 Ω is connected, and the load 2 of 10 Ω is parallel connected at $t = 2$ s. In Fig. 5.8a–c, the voltage amplitude and frequency are restored after load step change. Besides, as the m_p increases from 0.0001 to 0.05, the dynamic response of inverters is increasingly fast from Fig. 5.8a–c, where the response time is deceased from 1.6 to 1.3 s, since large droop control coefficients correspond to deep degree of adjustment, thereby the regulation time is shorter. Moreover, the drop depth of voltage amplitude is increasingly reduced from about 0.7 to 0.4 V. However, the drop depth of frequency is continuously increased from about 0.02 to 0.4 Hz

In Fig. 5.9, the condition of load step change is consistent with that of Fig. 5.8, and the voltage amplitude and frequency can be restored to the rated value after load step change. From Fig. 5.9a–c, the dynamic response time of the system is increasingly decreased from 0.7 to 0.5 s as n_q enlarges from 0.0001 to 0.05, which verifies the correctness of the theoretical analysis. Moreover, the drop depth of voltage amplitude is increased from 0.0001 to 0.05 V, and the drop depth of frequency is almost unchanged.

In Fig. 5.10, the condition of load step change is consistent with that of Fig. 5.8. From Fig. 5.10a, b, the frequency bandwidth f_{BW} of BPF is increased from 3 to 5 Hz. As seen in Fig. 5.10, the deviations of voltage amplitude and frequency can be restored in the selected parameters range, and the dynamic response is increasingly fast as the f_{BW} enlarges from Fig. 5.10a–c, where the response time is decreased from 1.3 to 0.9 s. In addition, the drop depth of voltage amplitude is continuously reduced from about 0.7 to 0.5 V, and the drop depth of frequency is diminished from about 0.06 to 0.004 Hz. Therefore, a relatively large bandwidth of BPF in the varied range should be selected for fast dynamic response and low drop depth.

In Fig. 5.11, different initial load 1 are connected before $t = 2$ s, and the load 2 is parallel connected at $t = 2$ s. As shown in Fig. 5.11a, b, the dynamic response time is continuously reduced from 0.8 to 0.6 s as load resistance increases. In addition, the drop depth of voltage amplitude is increasingly diminished from 0.01 to 0.0008 V, and the drop depth of frequency is increasingly reduced from 0.06 to 0.004 Hz. Hence, the simulation results demonstrate that a slightly large load resistance is more suitable for parameters selection, which is consistent with the eigenvalue analysis.

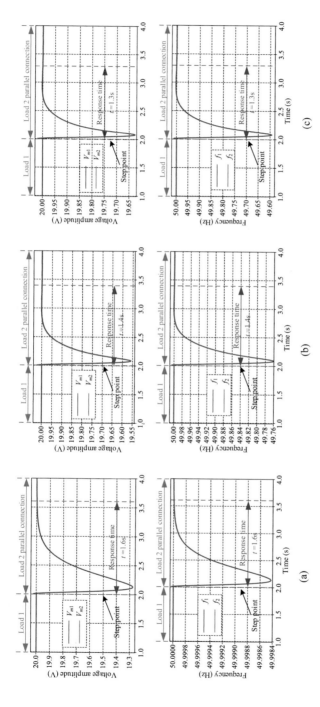

Fig. 5.8 Output voltage amplitude and frequency of inverters under different active droop control coefficients. (**a**) $m_p = 0.0001$, $f_{ch} = 1$ Hz. (**b**) $m_p = 0.025$, $f_{ch} = 3$ Hz. (**c**) $m_p = 0.05$, $f_{ch} = 4$ Hz

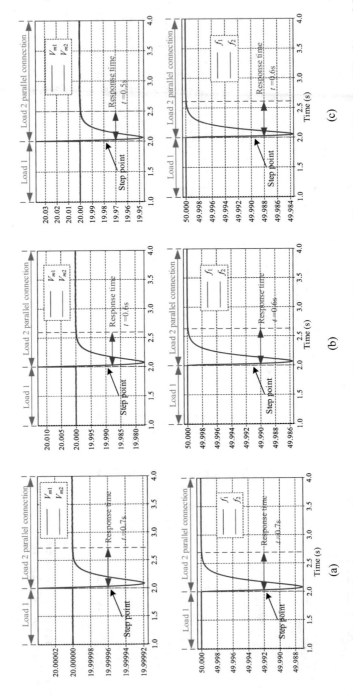

Fig. 5.9 Output voltage amplitude and frequency of inverters under different reactive droop control coefficients. (**a**) $n_q = 0.0001$, $f_{cl} = 3$ Hz. (**b**) $n_q = 0.025$, $f_{cl} = 4$ Hz. (**c**) $n_q = 0.05$, $f_{cl} = 6$ Hz

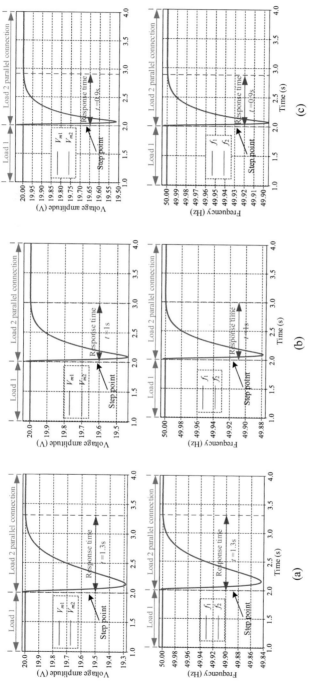

Fig. 5.10 Output voltage amplitude and frequency of inverters. (**a**) $f_{ch} = 1$ Hz, $f_{cl} = 4$ Hz. (**b**) $f_{ch} = 2$ Hz, $f_{cl} = 6$ Hz. (**c**) $f_{ch} = 3$ Hz, $f_{cl} = 8$ Hz

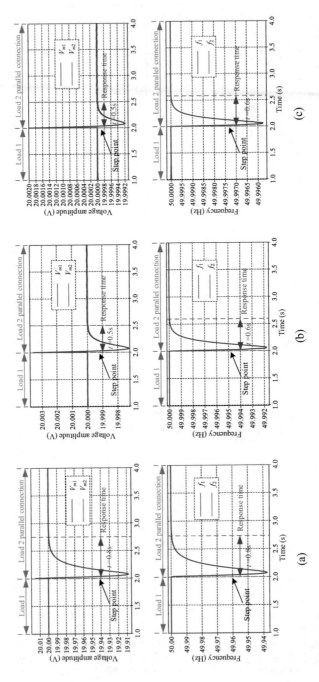

Fig. 5.11 Output voltage amplitude and frequency of each inverter. (**a**) Load 1 = 2 Ω, Load 2 = 10 Ω. (**b**) Load 1 = 20 Ω, Load 2 = 20 Ω. (**c**) Load 1 = 40 Ω, Load 2 = 40 Ω

5.5   Simulation Results from EMTP

Next, the simulation results obtained from EMTP software are presented, as a supplement to the previous section to study the comparative analysis between the conventional droop control and the BPF-based improved droop control scheme applied to the single phase islanded microgrid.

5.5.1   Performance of Conventional Droop Control Scheme

Figure 5.12 shows the simulation results when the conventional droop control scheme is utilized under identical power-stage parameter conditions, where $f_{cl} = 5$ Hz and $m_p = n_q = 0.05$. The controller is enabled at $t = 0.02$ s, and a transient increase of another 2 Ω load is parallel applied to the 10 Ω load at $t = 0.3$ s. Figure 5.12a shows the reference voltage amplitude and output voltage of each inverter, it can be observed that the output voltage undergoes a transient decrease when the 2 Ω load is abruptly applied. Figure 5.12b shows the reference frequency of each inverter, it can be observed that the frequency amplitude undergoes a transient decrease of 0.8 Hz when the 2 Ω load is abruptly applied. Figure 5.12c shows the active and reactive power of each inverter, which are identical to each other due to the identical power-stage and control parameters. Figure 5.12d shows the tracking performance of the first inverter, it can be observed that the output voltage across the filter capacitor tracks the reference voltage with high accuracy in the steady-state and dynamic process.

Figure 5.13 shows the simulation results when the conventional droop control scheme is utilized under identical power-stage parameter conditions, where $f_{cl} = 5$ Hz and $m_p = n_q = 0.01$. The system parameters and load dynamics are consistent with the previous case. It can be observed from Fig. 5.13 that the obtained results mimic the case in Fig. 5.12. However, due to a reduction of the droop coefficients, the voltage amplitude and reference frequency for each inverter unit undergoes a smaller transient disturbance when a transient increase of load is applied.

Figure 5.14 shows the simulation results when the conventional droop control scheme is utilized under non-identical power-stage parameter conditions, where the line impedance of inverter 2 is 20% higher than the nominal value, and $f_{cl} = 5$ Hz, $m_p = n_q = 0.05$. It can be observed from Fig. 5.14 that the obtained results mimic the case in Fig. 5.12. However, the voltage amplitude, reference frequency, active and reactive power for each inverter unit show non-identical waveforms during dynamic process when a transient increase of load is applied.

Figure 5.15 shows the simulation results when the conventional droop control scheme is utilized under non-identical power-stage parameter conditions, where the line impedance of inverter 2 is 20% higher than the nominal value, and $f_{cl} = 5$ Hz, $m_p = n_q = 0.01$. It can be observed from Fig. 5.14 that the obtained results mimic the case in Fig. 5.13. However, the voltage amplitude, reference frequency, active and reactive power for each inverter unit show non-identical waveforms during dynamic process when a transient increase of load is applied.

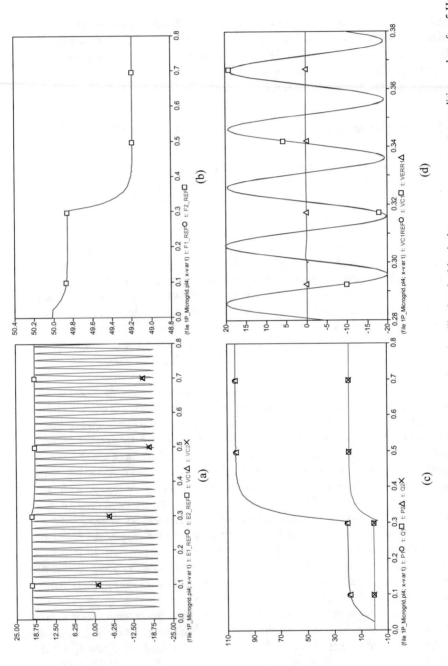

Fig. 5.12 Simulation results when the conventional droop control scheme is utilized under identical power-stage parameter conditions, where $f_{cl} = 5$ Hz and $m_p = n_q = 0.05$. (**a**) reference voltage amplitude and output voltage of each inverter; (**b**) reference frequency of each inverter; (**c**) the active and reactive power

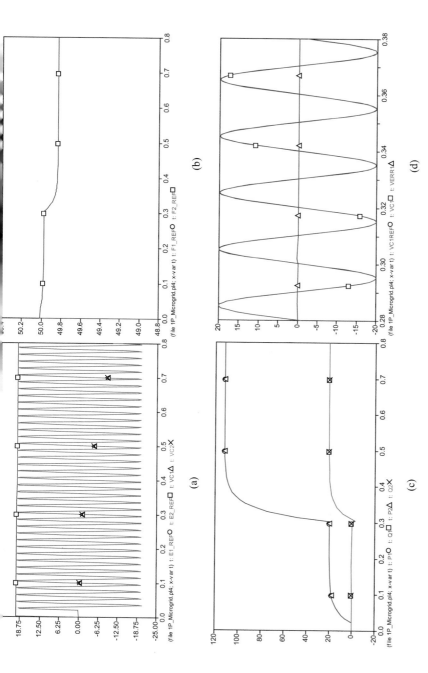

Fig. 5.13 Simulation results when the conventional droop control scheme is utilized under identical power-stage parameter conditions, where $f_{cl} = 5$ Hz and $m_p = n_q = 0.01$. (**a**) reference voltage amplitude and output voltage of each inverter; (**b**) reference frequency of each inverter; (**c**) the active and reactive power of each inverter; (**d**) the tracking performance evaluation of the first inverter

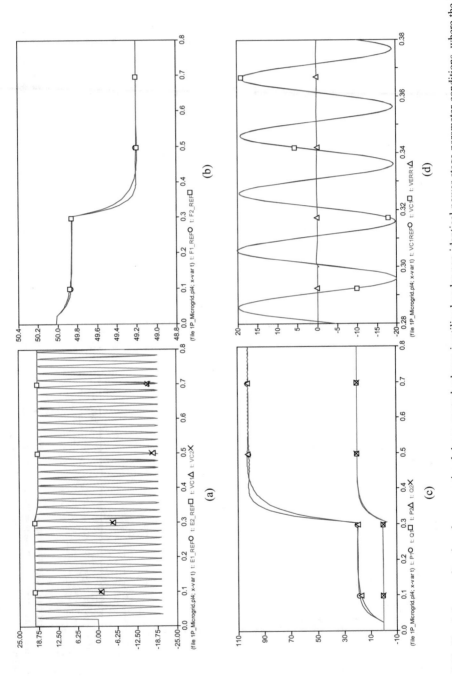

Fig. 5.14 Simulation results when the conventional droop control scheme is utilized under non-identical power-stage parameter conditions, where the line impedance of inverter 2 is 20% higher than the nominal value, and $f_{cd} = 5$ Hz, $m_p = n_q = 0.05$. (**a**) reference voltage amplitude and output voltage of each inverter;

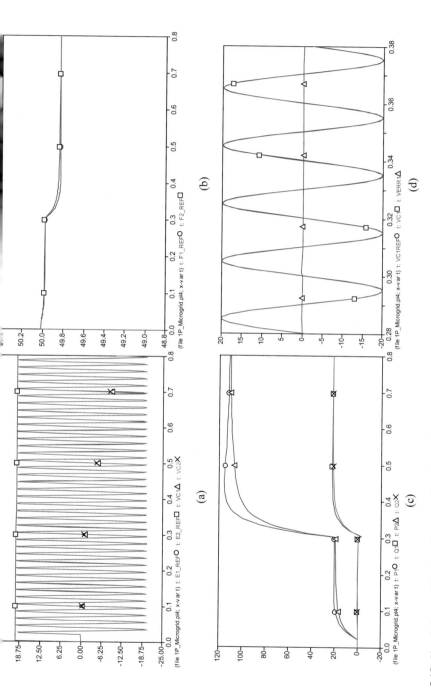

Fig. 5.15 Simulation results when the conventional droop control scheme is utilized under non-identical power-stage parameter conditions, where the line impedance of inverter 2 is 20% higher than the nominal value, and $f_{el} = 5$ Hz, $m_p = n_q = 0.01$. (**a**) reference voltage amplitude and output voltage of each inverter; (**b**) reference frequency of each inverter; (**c**) the active and reactive power of each inverter; (**d**) the tracking performance evaluation of the first inverter

Figure 5.16 shows the simulation results when the conventional droop control scheme is utilized under non-identical power-stage parameter conditions, where the line impedance of inverter 2 is 20% higher than the nominal value, and $f_{cl} = 10$ Hz, $m_p = n_q = 0.05$. It can be observed that, due to excessive high cutoff frequency of the low-pass filter, the reference voltage amplitude and frequency for each inverter unit undergoes divergence when a transient increase of load is applied. It can be inferred from Fig. 5.16 that the improper selection of the cutoff frequency in the low-pass filter in the droop controller would result in closed-loop instability of the system.

Figure 5.17 shows the simulation results when the conventional droop control scheme is utilized under non-identical power-stage parameter conditions, where the line impedance of inverter 2 is 20% higher than the nominal value, and $f_{cl} = 10$ Hz, $m_p = n_q = 0.01$. It shows in Fig. 5.17 that closed-loop stability is ensured when the droop coefficients are reduced, as compared to the case illustrated in Fig. 5.16. Due to the non-identical power-stage parameters, some difference can be observed during the transient process in the reference frequency, active and reactive power in each inverter.

On the other hand, in order to check the system robustness under the variation of filter capacitance. The deviation of 30% from the nominal value is studied, as shown in Figs. 5.18 and 5.19, respectively. It can be observed that, under both scenarios, the reference frequency and active and reactive power waveforms are almost identical to each other. Hence, it can be inferred that the conventional droop control scheme is not sensitive to the variation of filter capacitance.

5.5.2 Performance of BPF-Based Droop Control Scheme

From the simulation results in the previous subsection, it can be concluded that the conventional droop control scheme shows the shortcoming of steady-state error in the output voltage amplitude and frequency after a sudden increase of load is applied. The proposed BPF-based improved droop control scheme overcomes this shortcoming and eliminates the steady-state error, which mimics the secondary controller in the hierarchical control architecture.

Figure 5.20 shows the simulation results when the BPF-based droop control scheme is utilized under identical power-stage parameter conditions, where $f_{cl} = 5$ Hz, $f_{ch} = 2$ Hz and $m_p = n_q = 0.05$. Compared to the case studies in the previous subsection, it can be noticed that the output voltage amplitude and frequency of each inverter recovers to the nominal value in the steady state condition. Figure 5.21 shows the simulation results when the BPF-based droop control scheme is utilized under identical power-stage parameter conditions, where $f_{cl} = 5$ Hz, $f_{ch} = 2$ Hz and $m_p = n_q = 0.01$. It shows that the simulation results are almost identical with the case in Fig. 5.20. However, the transient amplitude and frequency dip in output voltage of each inverter is reduced due to a smaller droop coefficient.

Figure 5.22 shows the simulation results when the conventional droop control scheme is utilized under non-identical power-stage parameter conditions, where the line inductance of inverter 2 is 30% smaller than the nominal value, and $f_{cl} = 5$ Hz, $f_{ch} = 2$ Hz and $m_p = n_q = 0.05$. Figure 5.23 shows the simulation results when the

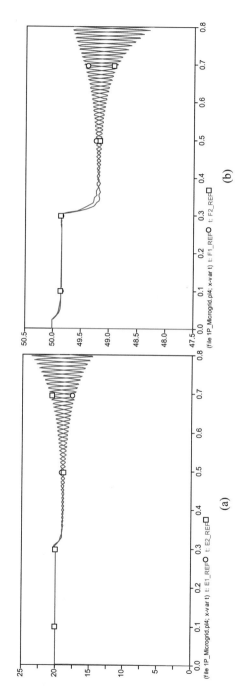

Fig. 5.16 Simulation results when the conventional droop control scheme is utilized under non-identical power-stage parameter conditions, where the line impedance of inverter 2 is 20% higher than the nominal value, and $f_{cl} = 10$ Hz, $m_p = n_q = 0.05$. (**a**) reference voltage amplitude of each inverter; (**b**) reference frequency of each inverter

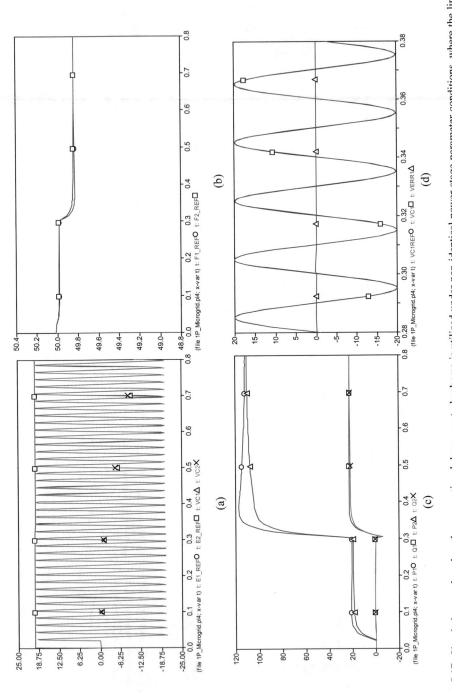

Fig. 5.17 Simulation results when the conventional droop control scheme is utilized under non-identical power-stage parameter conditions, where the line impedance of inverter 2 is 20% higher than the nominal value, and $f_{cl} = 10$ Hz, $m_p = n_q = 0.01$. (**a**) reference voltage amplitude and output voltage of each

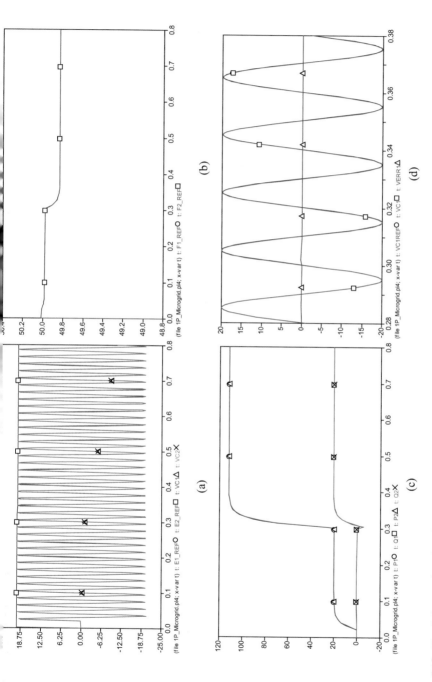

Fig. 5.18 Simulation results when the conventional droop control scheme is utilized under non-identical power-stage parameter conditions, where the filter capacitance of inverter 2 is 30% smaller than the nominal value, and $f_{cl} = 10$ Hz, $m_p = n_q = 0.01$. (**a**) reference voltage amplitude and output voltage of each inverter; (**b**) reference frequency of each inverter; (**c**) the active and reactive power of each inverter; (**d**) the tracking performance evaluation of the first inverter

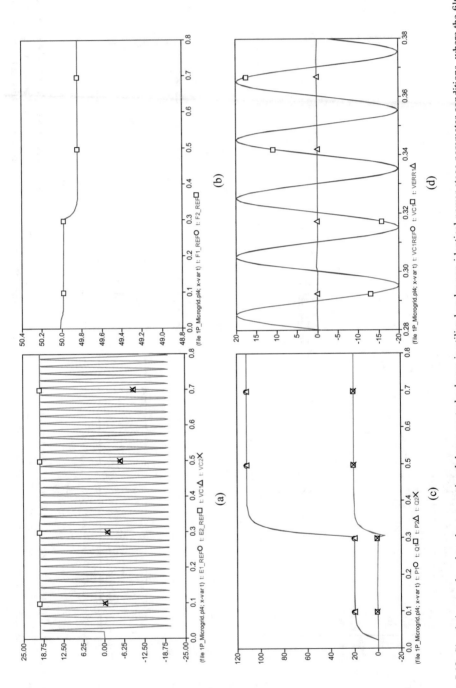

Fig. 5.19 Simulation results when the conventional droop control scheme is utilized under non-identical power-stage parameter conditions, where the filter capacitance of inverter 2 is 30% higher than the nominal value, and $f_{cl} = 10$ Hz, $m_p = n_q = 0.01$. (**a**) reference voltage amplitude and output voltage of each

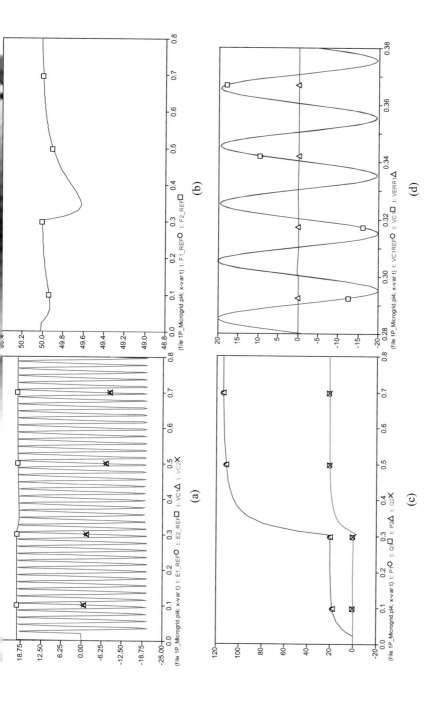

Fig. 5.20 Simulation results when the BPF-based droop control scheme is utilized under identical power-stage parameter conditions, where f_{cl} = 5 Hz, f_{ch} = 2 Hz and $m_p = n_q = 0.05$. (**a**) reference voltage amplitude and output voltage of each inverter; (**b**) reference frequency of each inverter; (**c**) the active and reactive power of each inverter; (**d**) the tracking performance evaluation of the first inverter

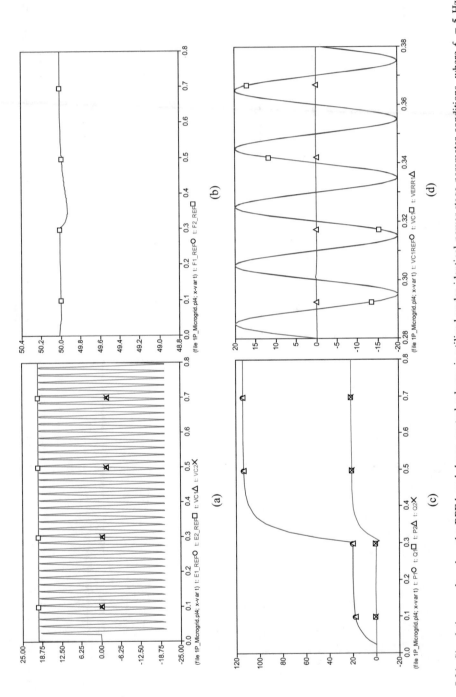

Fig. 5.21 Simulation results when the BPF-based droop control scheme is utilized under identical power-stage parameter conditions, where $f_{cd} = 5$ Hz, $f_{ch} = 2$ Hz and $m_p = n_q = 0.01$. (**a**) reference voltage amplitude and output voltage of each inverter; (**b**) reference frequency of each inverter; (**c**) the active and

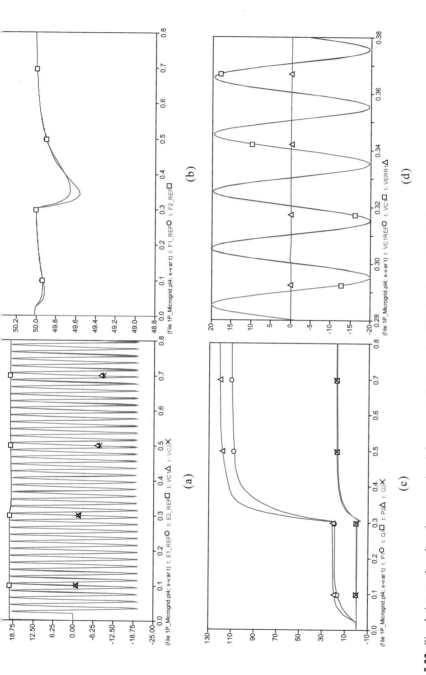

Fig. 5.22 Simulation results when the conventional droop control scheme is utilized under non-identical power-stage parameter conditions, where the line inductance of inverter 2 is 30% smaller than the nominal value, and $f_{cl} = 5$ Hz, $f_{ch} = 2$ Hz and $m_p = n_q = 0.05$. (**a**) reference voltage amplitude and output voltage of each inverter; (**b**) reference frequency of each inverter; (**c**) the active and reactive power of each inverter; (**d**) the tracking performance evaluation of the first inverter

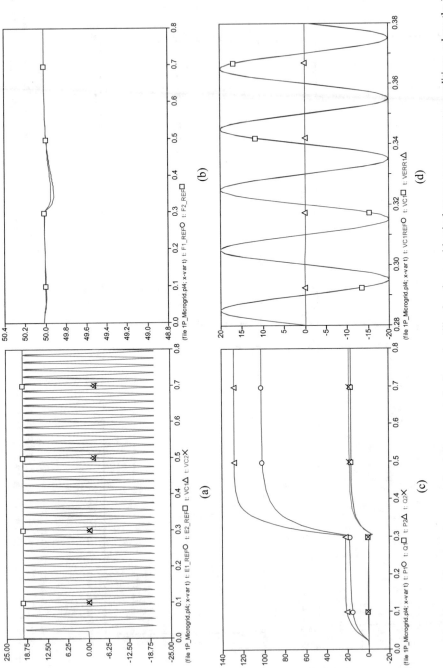

Fig. 5.23 Simulation results when the conventional droop control scheme is utilized under non-identical power-stage parameter conditions, where the line inductance of inverter 2 is 30% smaller than the nominal value, and $f_{cl} = 5$ Hz, $f_{ch} = 2$ Hz and $m_p = n_q = 0.01$. (a) reference voltage amplitude and output voltage

conventional droop control scheme is utilized under non-identical power-stage parameter conditions, where the line inductance of inverter 2 is 30% smaller than the nominal value, and $f_{cl} = 5$ Hz, $f_{ch} = 2$ Hz and $m_p = n_q = 0.01$. It shows that a higher droop coefficient result in a higher voltage amplitude and frequency dip in the output voltage of each inverter. However, the active power sharing accuracy can be improved, while the reactive power sharing accuracy is less sensitive to the droop coefficients.

Next, the effect of filter capacitance deviations on system performance is considered. Figure 5.24. shows the simulation results when the conventional droop control scheme is utilized under non-identical power-stage parameter conditions, where the filter capacitance of inverter 2 is 30% smaller than the nominal value, and $f_{cl} = 5$ Hz, $f_{ch} = 2$ Hz and $m_p = n_q = 0.05$. Figure 5.25 shows the simulation results when the conventional droop control scheme is utilized under non-identical power-stage parameter conditions, where the filter capacitance of inverter 2 is 30% higher than the nominal value, and $f_{cl} = 5$ Hz, $f_{ch} = 2$ Hz and $m_p = n_q = 0.05$. It shows that the reference amplitude and frequency in output voltage of each inverter are almost identical to each other in Figs. 5.24 and 5.25, which implies that the closed-loop control system is insensitive to the variation of filter capacitance when the BPF-based droop control scheme is applied.

5.6   Experimental Results and Discussions

In this subsection, a down-scaled experimental setup is utilized to further verify the feasibility of BPF-based droop control method in single-phase microgrid, as shown in Fig. 5.26. The DC-link voltage of two *LCL*-type full-bridge single-phase inverters is provided by programmable DC power supplies, where the load-side filter inductor is utilized to simulate the inductance X_t of long transmission lines. The control algorithm is implemented in TMS320F28335, and the load is linear resistance. Figures 5.27 and 5.28 show the experimental waveforms of conventional and BPF-based droop control methods, respectively, when the common load is abruptly changed. The waveforms from top to bottom are the capacitor voltages of inverter 1 and inverter 2, the load voltage, and the load current.

In Fig. 5.27a, the initial load 1 of the system is 10 Ω, then the load 2 of 10 Ω is sequentially parallel-connected with the initial load. The dotted lines in Fig. 5.27 represent the load step change instant. Consequently, the load voltage is significantly dropped, and it cannot be restored for a while. In Fig. 5.27b, the initial load of system is paralleled by two 10 Ω resistances, and the load 2 is subsequently disconnected. As shown in Fig. 5.27b, the load voltage is obviously rising after load step change, and it is also deviated from the reference value. Therefore, the load voltage error caused by the load step change cannot be well restored by the conventional droop control strategy.

In Fig. 5.28a, the load voltage and capacitor voltage of two inverters are still maintained at the rated value after incorporating a 10 Ω resistance, and the load

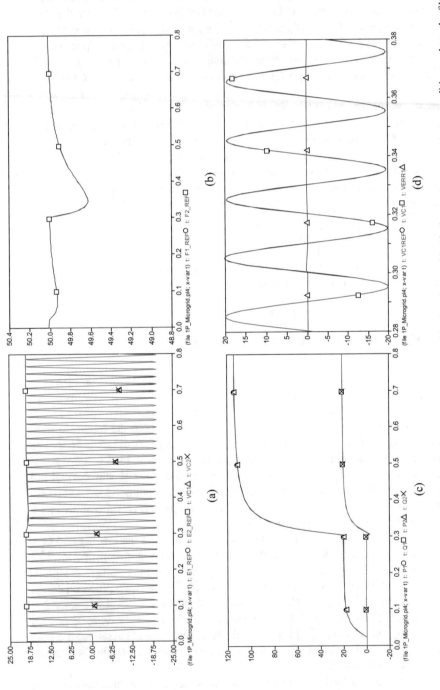

Fig. 5.24 Simulation results when the conventional droop control scheme is utilized under non-identical power-stage parameter conditions, where the filter capacitance of inverter 2 is 30% smaller than the nominal value, and $f_{cl} = 5$ Hz, $f_{ch} = 2$ Hz and $m_p = n_q = 0.05$. (**a**) reference voltage amplitude and output voltage of each inverter; (**b**) reference frequencies of each inverter; (**c**) the active and reactive power of each inverter; (**d**) the tracking performance evaluation of the first

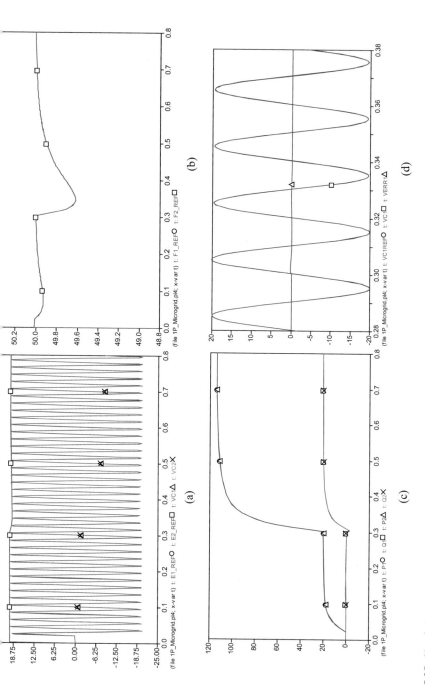

Fig. 5.25 Simulation results when the conventional droop control scheme is utilized under non-identical power-stage parameter conditions, where the filter capacitance of inverter 2 is 30% higher than the nominal value, and $f_{cl} = 5$ Hz, $f_{ch} = 2$ Hz and $m_p = n_q = 0.05$. (**a**) reference voltage amplitude and output voltage of each inverter; (**b**) reference frequency of each inverter; (**c**) the active and reactive power of each inverter; (**d**) the tracking performance evaluation of the first inverter

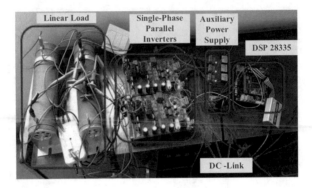

Fig. 5.26 Experimental setup of two-parallel single-phase inverters based islanded microgrid

current is evidently enlarged. Conversely, the load current is obviously reduced after load 2 disconnected in Fig. 5.28b, with the invariable load voltage. Hence, the feasibility of the BPF-based droop control scheme in the islanded single-phase microgrid is confirmed by comparing the experimental results shown in Figs. 5.27 and 5.28. After the load is changed, the output voltage deviation of the inverter can be restored to the nominal value by applying the BPF-based droop control method.

5.7   Conclusions

In this chapter, the BPF-based droop control is first extended to single-phase microgrid with SRF voltage control loop to achieve deviation restoration, which is universal and scalable both in the single-phase and three-phase microgrid. More importantly, any communication lines and extra control loops are no longer required in comparison with secondary control, which improves the reliable operation of system. Besides, the reduced-order state-space model of the system is subsequently derived by neglecting the internal fast states (inner control loops and LC filters) of inverters.

The small-signal stability of system is analyzed in case of shifted system parameters based on the eigenvalues loci of system matrix, which shows good stability, robustness and overdamped characteristics under a large variation range of control parameters. Then, the simulation results obtained from PLECS and EMTP software are provided to verify the effectiveness of eigenvalue-based theoretical analysis, the effect of controller parameters and power-stage parameter variations are also tested to study the system robustness. Finally, the feasibility of the BPF-based improved droop control scheme in single-phase microgrid controlled in hybrid reference frame is further validated by the experimental results.

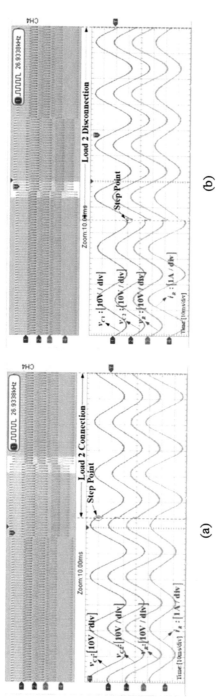

Fig. 5.27 Experimental waveforms of conventional droop control strategy. (**a**) Load 2 connection. (**b**) Load 2 disconnection

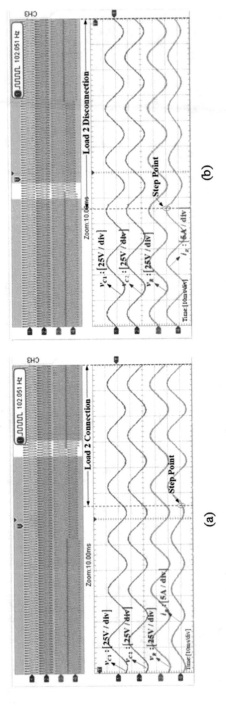

Fig. 5.28 Experimental waveforms of BPF-based droop control strategy. (**a**) Load 2 connection. (**b**) Load 2 disconnection

Appendix

The state vector is defined as

$$\Delta \mathbf{X}_B = [\Delta V_{m1e}, \Delta V_{m2e}, \Delta \delta_{1e}, \Delta \delta_{2e}, \Delta \omega_{f1e}, \Delta \omega_{f2e}, \Delta P_{1mese}, \Delta P_{2mese},$$
$$\Delta Q_{1mese}, \Delta Q_{2mese}]^T \tag{A.1}$$

The system state matrix is derived as

$$
A = \begin{bmatrix}
a_{b1,1} & a_{b1,2} & a_{b1,3} & a_{b1,4} & 0 & 0 & 0 & 0 & n_q\omega_{cl} & 0 \\
a_{b2,1} & a_{b2,2} & a_{b2,3} & a_{b2,4} & 0 & 0 & 0 & 0 & 0 & n_q\omega_{cl} \\
0 & 0 & 0 & 0 & 1 & 0 & 0 & 0 & 0 & 0 \\
0 & 0 & 0 & 0 & 0 & 1 & 0 & 0 & 0 & 0 \\
a_{b5,1} & a_{b5,2} & a_{b5,3} & a_{b5,4} & -\omega_{ch} & 0 & m_p\omega_{cl} & 0 & 0 & 0 \\
a_{b6,1} & a_{b6,2} & a_{b6,3} & a_{b6,4} & 0 & -\omega_{ch} & 0 & m_p\omega_{cl} & 0 & 0 \\
a_{b7,1} & a_{b7,2} & a_{b7,3} & a_{b7,4} & 0 & 0 & -\omega_{cl} & 0 & 0 & 0 \\
a_{b8,1} & a_{b8,2} & a_{b8,3} & a_{b8,4} & 0 & 0 & 0 & -\omega_{cl} & 0 & 0 \\
a_{b9,1} & a_{b9,2} & a_{b9,3} & a_{b9,4} & 0 & 0 & 0 & 0 & -\omega_{cl} & 0 \\
a_{b10,1} & a_{b10,2} & a_{b10,3} & a_{b10,4} & 0 & 0 & 0 & 0 & 0 & -\omega_{cl}
\end{bmatrix} \tag{A.2}
$$

The elements in the matrix $\mathbf{A}$ is presented as:

$$a_{b1,1} = -\left[n_q\omega_{cl} \left[\frac{V_{mle}}{|Z_t|}\sin\theta_t - \frac{V_{mle}}{|Z_t|} \cdot \frac{\cos(\delta_{1e}+\theta)\sin(\delta_{1e}+\theta_t)}{2\cos\theta + |Z|} \right. \right.$$
$$\left. \left. - \frac{V_{m2e}}{2|Z_t|} \cdot \frac{\cos(\delta_{2e}+\theta)\sin(\delta_{1e}+\theta_t)}{2\cos\theta + |Z|} \right] + \omega_{ch} \right] \tag{A.3}$$

$$a_{b1,2} = n_q\omega_{cl} \left[\frac{V_{mle}}{2|Z_t|} \cdot \frac{\cos(\delta_{2e}+\theta)\sin(\delta_{1e}+\theta_t)}{2\cos\theta + |Z|} \right] \tag{A.4}$$

$$a_{b1,3} = -n_q\omega_{cl} \left[\frac{V_{mle}^2}{2|Z_t|} \cdot \frac{\sin(\delta_{1e}+\theta)\sin(\delta_{1e}+\theta_t)}{2\cos\theta + |Z|} \right.$$
$$\left. - \frac{V_{mle}^2}{2|Z_t|} \cdot \frac{\cos(\delta_{1e}+\theta)\cos(\delta_{1e}+\theta_t)}{2\cos\theta + |Z|} - \frac{V_{mle}V_{m2e}}{2|Z_t|} \cdot \frac{\cos(\delta_{2e}+\theta)\cos(\delta_{1e}+\theta_t)}{2\cos\theta + |Z|} \right] \tag{A.5}$$

$$a_{b1,4} = -n_q\omega_{cl} \left[\frac{V_{mle}V_{m2e}}{2|Z_t|} \cdot \frac{\sin(\delta_{2e}+\theta)\sin(\delta_{1e}+\theta_t)}{2\cos\theta + |Z|} \right] \tag{A.6}$$

$$a_{b2,1} = n_q \omega_{cl} \left[\frac{V_{m2e}}{2|Z_t|} \cdot \frac{\cos(\delta_{1e} + \theta)\sin(\delta_{2e} + \theta_t)}{2\cos\theta + |Z|} \right] \tag{A.7}$$

$$a_{b2,2} = - \left[n_q \omega_{cl} \left[\frac{V_{m2e}}{|Z_t|} \sin\theta_t - \frac{V_{m1e}}{2|Z_t|} \cdot \frac{\cos(\delta_{1e} + \theta)\sin(\delta_{2e} + \theta_t)}{2\cos\theta + |Z|} \right. \right.$$
$$\left. \left. - \frac{V_{m2e}}{|Z_t|} \cdot \frac{\cos(\delta_{2e} + \theta)\sin(\delta_{2e} + \theta_t)}{2\cos\theta + |Z|} \right] + \omega_{ch} \right] \tag{A.8}$$

$$a_{b2,3} = -n_q \omega_{cl} \left[\frac{V_{m2e} V_{m1e}}{2|Z_t|} \cdot \frac{\sin(\delta_{1e} + \theta)\sin(\delta_{2e} + \theta_t)}{2\cos\theta + |Z|} \right] \tag{A.9}$$

$$a_{b2,4} = -n_q \omega_{cl} \left[\frac{V_{m2e}^2}{2|Z_t|} \cdot \frac{\sin(\delta_{2e} + \theta)\sin(\delta_{2e} + \theta_t)}{2\cos\theta + |Z|} \right.$$
$$\left. - \frac{V_{m2e} V_{m1e}}{2|Z_t|} \cdot \frac{\cos(\delta_{1e} + \theta)\cos(\delta_{2e} + \theta_t)}{2\cos\theta + |Z|} - \frac{V_{m2e}^2}{2|Z_t|} \cdot \frac{\cos(\delta_{2e} + \theta)\cos(\delta_{2e} + \theta_t)}{2\cos\theta + |Z|)} \right] \tag{A.10}$$

$$a_{b5,1} = -m_p \omega_{cl} \left[\frac{V_{m1e}}{|Z_t|} \cos\theta_t - \frac{V_{m1e}}{|Z_t|} \cdot \frac{\cos(\delta_{1e} + \theta)\cos(\delta_{1e} + \theta_t)}{2\cos\theta + |Z|} \right.$$
$$\left. - \frac{V_{m2e}}{2|Z_t|} \cdot \frac{\cos(\delta_{2e} + \theta)\cos(\delta_{1e} + \theta_t)}{2\cos\theta + |Z|} \right] \tag{A.11}$$

$$a_{b5,2} = m_p \omega_{cl} \left[\frac{V_{m1e}}{2|Z_t|} \cdot \frac{\cos(\delta_{2e} + \theta)\cos(\delta_{1e} + \theta_t)}{2\cos\theta + |Z|} \right] \tag{A.12}$$

$$a_{b5,3} = -m_p \omega_{cl} \left[\frac{V_{m1e}^2}{2|Z_t|} \cdot \frac{\sin(\delta_{1e} + \theta)\cos(\delta_{1e} + \theta_t)}{2\cos\theta + |Z|} \right.$$
$$\left. + \frac{V_{m1e}^2}{2|Z_t|} \cdot \frac{\cos(\delta_{1e} + \theta)\sin(\delta_{1e} + \theta_t)}{2\cos\theta + |Z|} + \frac{V_{m1e} V_{m2e}}{2|Z_t|} \cdot \frac{\cos(\delta_{2e} + \theta)\sin(\delta_{1e} + \theta_t)}{2\cos\theta + |Z|} \right] \tag{A.13}$$

$$a_{b5,4} = -m_p \omega_{cl} \left[\frac{V_{m1e} V_{m2e}}{2|Z_t|} \cdot \frac{\sin(\delta_{2e} + \theta)\cos(\delta_{1e} + \theta_t)}{2\cos\theta + |Z|} \right] \tag{A.14}$$

$$a_{b6,1} = m_p \omega_{cl} \left[\frac{V_{m2e}}{2|Z_t|} \cdot \frac{\cos(\delta_{1e} + \theta)\cos(\delta_{2e} + \theta_t)}{2\cos\theta + |Z|} \right] \qquad (A.15)$$

$$a_{b6,2} = -m_p \omega_{cl} \left[\frac{V_{m2e}}{|Z_t|}\cos\theta_t - \frac{V_{m1e}}{2|Z_t|} \cdot \frac{\cos(\delta_{1e} + \theta)\cos(\delta_{2e} + \theta_t)}{2\cos\theta + |Z|} \right. $$
$$\left. - \frac{V_{m2e}}{|Z_t|} \cdot \frac{\cos(\delta_{2e} + \theta)\cos(\delta_{2e} + \theta_t)}{2\cos\theta + |Z|} \right] \qquad (A.16)$$

$$a_{b6,3} = -m_p \omega_{cl} \left[\frac{V_{m1e}V_{m2e}}{2|Z_t|} \cdot \frac{\sin(\delta_{1e} + \theta)\cos(\delta_{2e} + \theta_t)}{2\cos\theta + |Z|} \right] \qquad (A.17)$$

$$a_{b6,4} = -m_p \omega_{cl} \left[\frac{V_{m2e}^2}{2|Z_t|} \cdot \frac{\sin(\delta_{2e} + \theta)\cos(\delta_{2e} + \theta_t)}{2\cos\theta + |Z|} \right.$$
$$\left. + \frac{V_{m2e}V_{m1e}}{2|Z_t|} \cdot \frac{\cos(\delta_{1e} + \theta)\sin(\delta_{2e} + \theta_t)}{2\cos\theta + |Z|} + \frac{V_{m2e}^2}{2|Z_t|} \cdot \frac{\cos(\delta_{2e} + \theta)\sin(\delta_{2e} + \theta_t)}{2\cos\theta + |Z|} \right] \qquad (A.18)$$

$$a_{b7,1} = \omega_{cl} \left[\frac{V_{m1e}}{|Z_t|}\cos\theta_t - \frac{V_{m1e}}{|Z_t|} \cdot \frac{\cos(\delta_{1e} + \theta)\cos(\delta_{1e} + \theta_t)}{2\cos\theta + |Z|} \right.$$
$$\left. - \frac{V_{m2e}}{2|Z_t|} \cdot \frac{\cos(\delta_{2e} + \theta)\cos(\delta_{1e} + \theta_t)}{2\cos\theta + |Z|} \right] \qquad (A.19)$$

$$a_{b7,2} = -\omega_{cl} \left[\frac{V_{m1e}}{2|Z_t|} \cdot \frac{\cos(\delta_{2e} + \theta)\cos(\delta_{1e} + \theta_t)}{2\cos\theta + |Z|} \right] \qquad (A.20)$$

$$a_{b7,3} = \omega_{cl} \left[\frac{V_{m1e}^2}{2|Z_t|} \cdot \frac{\sin(\delta_{1e} + \theta)\cos(\delta_{1e} + \theta_t)}{2\cos\theta + |Z|} \right.$$
$$\left. + \frac{V_{m1e}^2}{2|Z_t|} \cdot \frac{\cos(\delta_{1e} + \theta)\sin(\delta_{1e} + \theta_t)}{2\cos\theta + |Z|} + \frac{V_{m1e}V_{m2e}}{2|Z_t|} \cdot \frac{\cos(\delta_{2e} + \theta)\sin(\delta_{1e} + \theta_t)}{2\cos\theta + |Z|} \right] \qquad (A.21)$$

$$a_{b7,4} = \omega_{cl} \left[\frac{V_{m1e}V_{m2e}}{2|Z_t|} \cdot \frac{\sin(\delta_{2e} + \theta)\cos(\delta_{1e} + \theta_t)}{2\cos\theta + |Z|} \right] \qquad (A.22)$$

$$a_{b8,1} = -\omega_{cl}\left[\frac{V_{m2e}}{2|Z_t|}\cdot\frac{\cos(\delta_{1e}+\theta)\cos(\delta_{2e}+\theta_t)}{2\cos\theta+|Z|}\right] \tag{A.23}$$

$$a_{b8,2} = \omega_{cl}\left[\frac{V_{m2e}}{|Z_t|}\cos\theta_t - \frac{V_{mle}}{2|Z_t|}\cdot\frac{\cos(\delta_{1e}+\theta)\cos(\delta_{2e}+\theta_t)}{2\cos\theta+|Z|}\right.$$
$$\left. -\frac{V_{m2e}}{|Z_t|}\cdot\frac{\cos(\delta_{2e}+\theta)\cos(\delta_{2e}+\theta_t)}{2\cos\theta+|Z|}\right] \tag{A.24}$$

$$a_{b8,3} = \omega_{cl}\left[\frac{V_{mle}V_{m2e}}{2|Z_t|}\cdot\frac{\sin(\delta_{1e}+\theta)\cos(\delta_{2e}+\theta_t)}{2\cos\theta+|Z|}\right] \tag{A.25}$$

$$a_{b8,4} = \omega_{cl}\left[\frac{V_{m2e}^2}{2|Z_t|}\cdot\frac{\sin(\delta_{2e}+\theta)\cos(\delta_{2e}+\theta_t)}{2\cos\theta+|Z|}\right.$$
$$+\frac{V_{m2e}V_{mle}}{2|Z_t|}\cdot\frac{\cos(\delta_{1e}+\theta)\sin(\delta_{2e}+\theta_t)}{2\cos\theta+|Z|}+\frac{V_{m2e}^2}{2|Z_t|}\cdot\frac{\cos(\delta_{2e}+\theta)\sin(\delta_{2e}+\theta_t)}{2\cos\theta+|Z|}\right] \tag{A.26}$$

$$a_{b9,1} = \omega_{cl}\left[\frac{V_{mle}}{|Z_t|}\sin\theta_t - \frac{V_{mle}}{|Z_t|}\cdot\frac{\cos(\delta_{1e}+\theta)\sin(\delta_{1e}+\theta_t)}{2\cos\theta+|Z|}\right.$$
$$\left. -\frac{V_{m2e}}{2|Z_t|}\cdot\frac{\cos(\delta_{2e}+\theta)\sin(\delta_{1e}+\theta_t)}{2\cos\theta+|Z|}\right] \tag{A.27}$$

$$a_{b9,2} = -\omega_{cl}\left[\frac{V_{mle}}{2|Z_t|}\cdot\frac{\cos(\delta_{2e}+\theta)\sin(\delta_{1e}+\theta_t)}{2\cos\theta+|Z|}\right] \tag{A.28}$$

$$a_{b9,3} = \omega_{cl}\left[\frac{V_{mle}^2}{2|Z_t|}\cdot\frac{\sin(\delta_{1e}+\theta)\sin(\delta_{1e}+\theta_t)}{2\cos\theta+|Z|}\right.$$
$$\left. -\frac{V_{mle}^2}{2|Z_t|}\cdot\frac{\cos(\delta_{1e}+\theta)\cos(\delta_{1e}+\theta_t)}{2\cos\theta+|Z|}-\frac{V_{mle}V_{m2e}}{2|Z_t|}\cdot\frac{\cos(\delta_{2e}+\theta)\cos(\delta_{1e}+\theta_t)}{2\cos\theta+|Z|}\right] \tag{A.29}$$

$$a_{b9,4} = \omega_{cl}\left[\frac{V_{mle}V_{m2e}}{2|Z_t|}\cdot\frac{\sin(\delta_{2e}+\theta)\sin(\delta_{1e}+\theta_t)}{2\cos\theta+|Z|}\right] \tag{A.30}$$

$$a_{b10,1} = -\omega_{cl} \left[\frac{V_{m2e}}{2|Z_t|} \cdot \frac{\cos(\delta_{1e}+\theta)\sin(\delta_{2e}+\theta_t)}{2\cos\theta+|Z|} \right] \tag{A.31}$$

$$a_{b10,2} = \omega_{cl} \left[\frac{V_{m2e}}{|Z_t|}\sin\theta_t - \frac{V_{m1e}}{2|Z_t|} \cdot \frac{\cos(\delta_{1e}+\theta)\sin(\delta_{2e}+\theta_t)}{2\cos\theta+|Z|} \right.$$
$$\left. - \frac{V_{m2e}}{|Z_t|} \cdot \frac{\cos(\delta_{2e}+\theta)\sin(\delta_{2e}+\theta_t)}{2\cos\theta+|Z|} \right] \tag{A.32}$$

$$a_{b10,3} = \omega_{cl} \left[\frac{V_{m2e}V_{m1e}}{2|Z_t|} \cdot \frac{\sin(\delta_{1e}+\theta)\sin(\delta_{2e}+\theta_t)}{2\cos\theta+|Z|} \right] \tag{A.33}$$

$$a_{b10,4} = \omega_{cl} \left[\frac{V_{m2e}^2}{2|Z_t|} \cdot \frac{\sin(\delta_{2e}+\theta)\sin(\delta_{2e}+\theta_t)}{2\cos\theta+|Z|} \right.$$
$$- \frac{V_{m2e}V_{m1e}}{2|Z_t|} \cdot \frac{\cos(\delta_{1e}+\theta)\cos(\delta_{2e}+\theta_t)}{2\cos\theta+|Z|} - \frac{V_{m2e}^2}{2|Z_t|} \cdot \frac{\cos(\delta_{2e}+\theta)\cos(\delta_{2e}+\theta_t)}{2\cos\theta+|Z|} \right] \tag{A.34}$$

$$\theta = \theta_{load} - \theta_t \tag{A.35}$$

$$Z = |Z_t| / |Z_{load}| \tag{A.36}$$

References

1. Guerrero, J. M., Vasquez, J. C., Matas, J., Vicuna, L. G. d., & Castilla, M. (2011). Hierarchical control of droop-controlled AC and DC microgrids—A general approach toward standardization. *IEEE Transactions on Industrial Electronics, 58*(1), 158–172.
2. Han, Y., Li, H., Shen, P., Coelho, E. A. A., & Guerrero, J. M. (2017). Review of active and reactive power sharing strategies in hierarchical controlled microgrids. *IEEE Transactions on Power Electronics, 32*(3), 2427–2451.
3. Meng, X., Liu, J., & Liu, Z. (2019). A generalized droop control for grid-supporting inverter based on comparison between traditional droop control and virtual synchronous generator control. *IEEE Transactions on Power Electronics, 34*(6), 5416–5438.
4. Golsorkhi, M. S., Shafiee, Q., Lu, D. D., & Guerrero, J. M. (2019). Distributed control of low-voltage resistive AC microgrids. *IEEE Transactions on Energy Conversion, 34*(2), 573–584.
5. Lou, G., Gu, W., Sheng, W., Song, X., & Gao, F. (2018). Distributed model predictive secondary voltage control of islanded microgrids with feedback linearization. *IEEE Access, 6*, 50169–50178.
6. Shen, X., Wang, H., Li, J., Su, Q., & Gao, L. (2019). Distributed secondary voltage control of islanded microgrids based on RBF-neural-network sliding-mode technique. *IEEE Access, 7*, 65616–65623.

7. Dehkordi, N. M., Baghaee, H. R., Sadati, N., & Guerrero, J. M. (2019). Distributed noise-resilient secondary voltage and frequency control for islanded microgrids. *IEEE Transactions on Smart Grid, 10*(4), 3780–3790.

8. Yazdanian, M., & Mehrizi-Sani, A. (2016). Washout filter-based power sharing. *IEEE Transactions on Smart Grid, 7*(2), 967–968.

9. Shan, Y., Hu, J., Liu, M., Zhu, J., & Guerrero, J. M. Model predictive voltage and power control of islanded PV-battery microgrids with washout filter based power sharing strategy. *IEEE Transactions on Power Electronics*, to be published. https://doi.org/10.1109/TPEL.2019.2930182

10. Han, Y., Li, H., Xu, L., Zhao, X., & Guerrero, J. M. (2018). Analysis of washout filter-based power sharing strategy——An equivalent secondary controller for islanded microgrid without LBC lines. *IEEE Transactions on Smart Grid, 9*(5), 4061–4076.

11. Rasheduzzaman, M., Mueller, J. A., & Kimball, J. W. (2015). Reduced-order small-signal model of microgrid systems. *IEEE Transactions on Sustainable Energy, 6*(4), 1292–1305.

12. Mariani, V., Vasca, F., Vásquez, J. C., & Guerrero, J. M. (2015). Model order reductions for stability analysis of islanded microgrids with droop control. *IEEE Transactions on Industrial Electronics, 62*(7), 4344–4354.

13. Nikolakakos, I. P., Zeineldin, H. H., El-Moursi, M. S., & Kirtley, J. L. (2018). Reduced-order model for inter-inverter oscillations in islanded droop-controlled microgrids. *IEEE Transactions on Smart Grid, 9*(5), 4953–4963.

14. Lenz, E., Pagano, D. J., & Pou, J. (2018). Bifurcation analysis of parallel-connected voltage-source inverters with constant power loads. *IEEE Transactions on Smart Grid, 9*(6), 5482–5493.

15. Shuai, Z., Peng, Y., Liu, X., Li, Z., Guerrero, J. M., & Shen, Z. J. (2019). Dynamic equivalent modeling for multi-microgrid based on structure preservation method. *IEEE Transactions on Smart Grid, 10*(4), 3929–3942.

16. Purba, V., Johnson, B. B., Rodriguez, M., Jafarpour, S., Bullo, F., & Dhople, S. (2019). Reduced-order aggregate model for parallel-connected single-phase inverters. *IEEE Transactions on Energy Conversion, 34*(2), 824–837.

17. Luo, L., & Dhople, S. V. (2014). Spatiotemporal model reduction of inverter-based islanded microgrids. *IEEE Transactions on Energy Conversion, 29*(4), 823–832.

18. Floriduz, A., Tucci, M., Riverso, S., & Ferrari-Trecate, G. Approximate Kron reduction methods for electrical networks with applications to plug-and-play control of AC islanded microgrids. *IEEE Transactions on Control Systems Technology*, to be published. https://doi.org/10.1109/TCST.2018.2863645

19. Yang, M., Wang, C., Han, Y., Xiong, J., & Yang, P. (2018). Small-signal stability analysis of the standalone single-phase parallel inverter with BPF-based droop control scheme. In *Proc. IEEE Southern Power Electron. Conf. (SPEC)* (pp. 773–780).

20. Han, Y., Yang, M., Yang, P., Xu, L., Fang, X., Zhang, K., & Blaabjerg, F. (2019). Reduced-order model for dynamic stability analysis of single-phase islanded microgrid with BPF-based droop control scheme. *IEEE Acess, 7*, 157859–157872.

21. Yu, K., Ai, Q., Wang, S., Ni, J., & Lv, T. (2016). Analysis and optimization of droop controller for microgrid system based on small-signal dynamic model. *IEEE Transactions on Smart Grid, 7*(2), 695–705.

Chapter 6
Enhanced Droop Control Strategy for Three-Phase Islanded Microgrid Without LBC Lines

In the hierarchical control framework, the concept of secondary control has been extensively studied for voltage and frequency restoration of islanded or grid-connected microgrid. However, the low-bandwidth communication (LBC) channels are needed to exchange information between the primary and secondary controllers, and the performance of the secondary control would degrade due to the uncertain communication delay and data drop-out in the LBC lines. To overcome this shortcoming, the washout filter-based power sharing method can be applied without using low-bandwidth communication lines and additional control loops. The application of this concept has been applied in islanded single-phase ac microgrid in the previous chapter; hence, this idea can be extended to three-phase islanded ac microgrid. In this chapter, the equivalence between secondary control scheme and washout filter-based power sharing strategy for islanded microgrid is demonstrated, and the generalized washout filter control scheme is derived. In addition, the physical meaning of control parameters of secondary controllers is also illustrated. Besides, a complete small-signal model of the generalized washout filter-based control method for islanded Microgrid (MG) system is built, which can be utilized to design the control parameters and analyze the stability of MG system. Moreover, the simulation results obtained from EMTP are given to illustrate the difference between the conventional droop control scheme and the washout filter-based improved droop control scheme, and hardware-in-the-loop results are also presented to show a comparative analysis under generic operating conditions. Finally, the experimental results from a reduced-scale prototype system are provided to confirm the validity and effectiveness of the derived equivalent control scheme for three-phase islanded MG.

© Springer Nature Switzerland AG 2022
Y. Han, *Modeling and Control of Power Electronic Converters for Microgrid Applications*, https://doi.org/10.1007/978-3-030-74513-4_6

6.1 Introduction

The accurate power sharing while maintaining stable regulation of the MG voltage magnitude and frequency is crucial in an islanded microgrid. In the existing literatures, the droop control methods are widely adopted for a large/medium system, which mimics the behavior of a synchronous generator with no need of critical communication, to achieve the power sharing requirement eliminating an external high bandwidth communication links among the Distributed Generation (DG) units. Since the frequency is a global variable, the active power can be properly shared using the droop control, but the frequency and voltage amplitude deviations are inevitable in the steady-state conditions. Moreover, the dynamic stability of the active power sharing controller is poor and the power sharing accuracy is sensitive to the feeder impedance [1–3].

In order to achieve a tradeoff between the voltage regulation, frequency restoration and power sharing in the islanded microgrid, a number of improved control methods have been proposed, which can be divided into improved droop control and improved secondary control methods. An improved virtual power-based control method with a unified rotation angle in the power transformation has been presented in [4], which can effectively realize power decoupling and then ensure system stability. However, since the voltage and current control loops, filters, and loads are not considered in the small-signal model of the presented control strategy, the analysis of stability of the system is incomplete. A fuzzy logic-based improved droop control method is presented to balance the state of charge of DG energy storage systems [5]. However, the disturbance of the feeders and loads are not considered. An improved droop control method was proposed in [6] to share the DG currents and restore the bus voltage simultaneously without a centralized secondary controller. However, the communication lines are needed in this control strategy and the communication delay cannot be ignored. In [7], a fuzzy approach for intelligent model-based droop control has been established to regulate the MG frequency and voltage amplitude simultaneously. However, it makes the control structure more complicated, and the influence of the unequal feeder impedance was not considered. In order to improve the dynamic performance of the MG, the derivative control can be applied into the droop controller. However, the derivative control may make the parallel DG system unstable, especially when the DG unit is under no-load conditions.

As the main trend methods, the distributed/centralized secondary control approach can be used to restore the voltage amplitude and frequency to the rated values. In [8], a consensus-based secondary control strategy is presented to achieve accurate active power sharing in islanded MG with sparse communication lines. In [9], a two-layer cooperative control strategy is presented to simultaneously control both the voltage/frequency as well as the active/reactive power flows, where only own and neighbors' information of each DG unit are required. The improved secondary control strategies, such as the algorithms based on graph theory, predictive control, and multi-agent system (MAS)-based control methods, are presented to enhance the dynamic stability and accuracy of the power sharing under changeable environmental conditions. However, the low bandwidth communication (LBC) lines are inevitable to be utilized in these improved secondary control methods, and the output correction signals sent to the primary control are always accompanied by time delay and the control signals might be different from the theoretical analysis, which degrades the performance of the microgrid.

In this chapter, the equivalence between secondary control and washout filter-based control strategy is demonstrated, and the generalized washout filter-based power sharing scheme can be derived. Additionally, the physical meaning of parameters of the secondary controllers is discussed, emphasizing that the proportional and integral (PI) coefficients of the secondary controller are utilized to constitute a band-pass filter (BPF). Furthermore, a complete small-signal model of the generalized washout filter-based control method, considering the power stages, voltage and current controllers, LCL filters, feeder and load impedances, is proposed to design the control parameters of the calculated equivalent control model, and analyze the stability of the system. The feasibility and effectiveness of the proposed approach is validated by the hardware-in-the-loop (HIL) results obtained from the three parallel DG units-based islanded microgrid under unequal feeder impedance and load/DG disturbance conditions using the dSPACE 1006 control platform. Moreover, a down-scaled hardware prototype of islanded microgrid using two parallel-connected three-phase inverters is built and the experimental results are presented for verification. And the future research trends for the hierarchical controlled islanded microgrid is also discussed.

The rest of this chapter is organized as follows. In Sect. 6.2, the review of the secondary and washout filter-based control methods is presented. The equivalence between the secondary control and washout filter-based methods is presented, and the generalized washout filter-based control scheme can be obtained. In Sect. 6.3, the detailed small-signal model of the generalized washout filter-based power sharing method is established to design the parameters in the derived control method and the stability of the islanded MG system is analyzed. The EMTP simulation results and the HIL test results of the three DG-based islanded MG system are presented to verify the feasibility of the presented method in Sections 6.4 and 6.5. In Sect. 6.6, the experimental results from a down-scaled hardware prototype of islanded microgrid is presented, which verifies the correctness and effectiveness of the obtained equivalent control model. Finally, the concluding remarks are given in Sect. 6.7.

6.2   Secondary and Washout Filter-Based Control Strategies

In an islanded microgrid, a secondary control performs the function to eliminate the frequency and voltage deviations caused by the droop control algorithm, and maintain the stability of voltage and frequency of microgrid simultaneously. Additionally, to eliminate the impact of time delay caused by LBC lines in secondary controllers, a washout filter-based power sharing strategy without any communication link has been presented in [10].

Figure 6.1 shows a power stage of a DG unit with secondary or washout filter-based controller for the interface inverter in an islanded mode. As depicted in Fig. 6.1, each DG unit can be connected to a predefined load or to the common bus directly, which can be considered as a subsystem of the MG. A brief description of secondary control and washout filter-based control strategy for the MG are outlined as follows [8–10].

6.2.1   The Secondary Control for Islanded AC Microgrid

In islanded MG, the foundation of control loops of each DG unit must be established to stabilize the network and achieve a good power sharing among the DG units. Therefore, the classical (P/f, Q/V) droop control scheme in large systems (high voltage) and medium systems (medium voltage) is introduced, which can be defined as [11]:

$$\begin{cases} \omega = \omega^* - m_p P \\ v = v^* - n_q Q \end{cases} \tag{6.1}$$

where ω and v represent the frequency and amplitude of the output voltage. ω^* and v^* are the rated angular frequency and voltage, respectively. P and Q are the measured average active and reactive powers through a low-pass filter, and m_p and n_q are the frequency and amplitude droop coefficients, respectively.

The droop controller is responsible for adjusting the frequency and the amplitude of the voltage reference according to the active and reactive powers (P and Q), and to achieve the active power sharing among multiple DG units. However, the deviations of the frequency and voltage amplitude are inevitable, and the dynamic stability of the active power sharing is poor with the disturbances of loads and feeder dynamics. Therefore, in order to solve the problems caused by the conventional droop control, a secondary control can be used to eliminate the frequency and voltage deviations, and improve the stability of the MG [12, 13].

Figure 6.1 depicts the details of a secondary control structure, which is realized by using low bandwidth communication (LBC) among the multiple DG units. The secondary control consists of a proportional–integral (PI) controller, and the frequency and amplitude restoration compensators can be derived as:

$$\begin{cases} \omega_{sec} = k_{p\omega} \left(\omega_{MG}^* - \omega \right) + k_{i\omega} \int \left(\omega_{MG}^* - \omega \right) dt \\ v_{sec} = k_{pE} \left(v_{MG}^* - v \right) + k_{iE} \int \left(v_{MG}^* - v \right) dt \end{cases} \tag{6.2}$$

where $k_{p\omega}$, $k_{i\omega}$, k_{pE}, and k_{iE} are the control parameters of the PI compensator of the frequency and voltage restoration control, respectively. The errors between measured angular frequency (ω) and reference angular frequency ($\omega* MG$) are

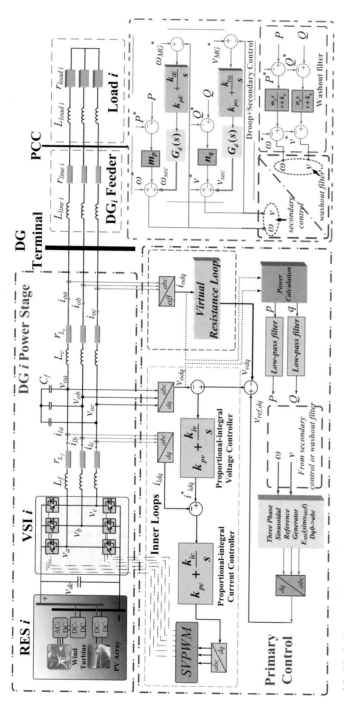

Fig. 6.1 Block diagram of the complete microgrid system including the secondary control or the washout filter-based control schemes

processed by the PI compensator and then sent the control signal ω_{sec} to all the DG units to restore the frequency of MG to the rated value. The control signal v_{sec} is also sent to primary control level to remove the voltage difference brought by the droop controller.

Notably, the centralized secondary control architecture requires each DG unit to communicate with a central controller, or requires all DGs to communicate with all others directly. Therefore, a MG will be unstable when the output frequency and voltage amplitude correction signals sent to primary control are different to the theoretical values, due to the low bandwidth communication (LBC) delays and data drop in the communication lines. Therefore, the secondary control for active power sharing should be further improved to get an accurate and robust active power sharing for the MGs, and decrease the impact of the LBC delays and data drop.

6.2.2   Washout Filter-Based Power Sharing Strategies for Islanded AC Microgrid

To eliminate the impact of time delay caused by the LBC lines and restore the frequency and voltage amplitude to the rated values simultaneously, a washout filter-based power sharing has been presented in [10] without any communication links and additional control loops as follows:

$$\begin{cases} \omega = \omega^* - \dfrac{m_p s}{s+k_p}\left(P-P^*\right) \\[4mm] v = v^* - \dfrac{n_q s}{s+k_q}\left(Q-Q^*\right) \end{cases} \tag{6.3}$$

where k_p and k_q are the control parameters of the washout filter. By using the control strategy as indicated in (6.3), voltage and frequency deviations can be prevented without the need for the secondary level control and extra controllers, where the droop coefficients are replaced by washout filters.

Notice that the washout filter-based control strategy is an equivalent secondary controller, which will be analyzed in next section. Moreover, the physical meaning of parameters of the secondary controllers will also be discussed in the next subsection.

6.2.3   Equivalence Between Secondary and Washout Filter-Based Controllers

Usually, frequency and voltage deviations from the nominal values due to the droop algorithm can be compensated by a secondary control. Referring to Fig. 6.1, the output voltage amplitude and frequency of the droop controller can be obtained as:

$$\begin{cases} \omega = \omega^* - m_p \left(\underbrace{p \cdot G_{LPF}(s) - P^*}_{P} \right) + \omega_{sec} \\ \\ v = v^* - n_q \left(\underbrace{q \cdot G_{LPF}(s) - Q^*}_{Q} \right) + v_{sec} \end{cases} \tag{6.4}$$

where p and q are instantaneous active and reactive powers, respectively. In the secondary voltage amplitude control loop, v_{sec} can be obtained as:

$$\begin{aligned} v_{sec} &= \left(v_{MG}^* - v \right) \cdot G_{v,sec}(s) \cdot G_d(s) \\ &= \left(v_{MG}^* - \left(v^* - n_q \left(q \cdot G_{LPF}(s) - Q^* \right) + v_{sec} \right) \right) \\ &\quad \cdot G_{v,sec}(s) \cdot G_d(s) \end{aligned} \tag{6.5}$$

where $G_d(s)$ is the transfer function of unknown LBC delay. The transfer function $G_{LPF}(s)$, $G_{v,sec}(s)$, and $G_{\omega,sec}(s)$ are defined as follows:

$$\begin{cases} G_{v,sec}(s) = k_{pE} + \dfrac{k_{iE}}{s} \\ \\ G_{\omega,sec}(s) = k_{p\omega} + \dfrac{k_{i\omega}}{s} \\ \\ G_{LPF}(s) = \dfrac{\omega_c}{s + \omega_c} \end{cases} \tag{6.6}$$

where ω_c is the cut-off frequency of the low-pass filter (LPF).

Moreover, the LBC delay is uncertain in the secondary controlled islanded microgrid, which may affect the stability of the system. Under ideal circumstances, the transfer function of LBC delay $G_d(s)$ is considered to be unity. The reference voltage of the microgrid $v*MG$ is set to be v^* and the reference angular frequency of MG $\omega*MG$ is set to be ω^*. Besides, the reference powers P^* and Q^* are set to be zero for islanded microgrid [14, 15].

Therefore, (6.5) can be simplified as:

$$v_{sec} = \dfrac{n_q}{\dfrac{1}{G_{v,sec}} + 1} \left(Q \cdot G_{LPF}(s) - Q^* \right) \tag{6.7}$$

Combining (6.4) and (6.7), the output voltage can be obtained as:

$$v = v^* - n_q \left(Q \cdot G_{LPF}(s) - Q^* \right) + \frac{n_q}{\dfrac{1}{G_{v,\sec}} + 1} \left(Q \cdot G_{LPF}(s) - Q^* \right)$$

$$= v^* - n_q \cdot \frac{\omega_c}{s + \omega_c} \cdot \frac{1}{G_{v,\sec} + 1} \cdot q \tag{6.8}$$

$$= v^* - \frac{n_q}{k_{pE} + 1} \cdot \underbrace{\frac{\omega_c}{s + \omega_c}}_{low-pass} \cdot \underbrace{\frac{s}{s + \dfrac{k_{iE}}{k_{pE} + 1}}}_{high-pass} \cdot q$$

$$\underbrace{\hspace{6cm}}_{band-pass\ filter}$$

Besides, the angular frequency can be derived as:

$$\omega = \omega^* - \frac{m_p}{k_{p\omega} + 1} \cdot \underbrace{\frac{\omega_c}{s + \omega_c}}_{low-pass} \cdot \underbrace{\frac{s}{s + \dfrac{k_{i\omega}}{k_{p\omega} + 1}}}_{high-pass} \cdot p \tag{6.9}$$

$$\underbrace{\hspace{6cm}}_{band-pass\ filter}$$

Therefore, from (6.8) and (6.9), a generalized washout filter-based power sharing strategy can be obtained. Note that a washout filter-based control strategy can be achieved when the conditions $k_{p\omega} = k_{pE} = 0$, $k_{i\omega} = k_p$, and $k_{iE} = k_q$ are satisfied. Moreover, it can be concluded that the washout filter-based power sharing strategy is intrinsically an ideal secondary control without communication delay.

As can be seen in (6.8) and (6.9), cut-off frequencies ω_{hE} and $\omega_{h\omega}$ of high-pass filter (HPF) are constituted by the proportional and integral coefficients of the secondary controller, where ω_{hE} and $\omega_{h\omega}$ are defined as:

$$\omega_{hE} = \frac{k_{iE}}{k_{pE} + 1}, \quad \omega_{h\omega} = \frac{k_{i\omega}}{k_{p\omega} + 1} \tag{6.10}$$

From (6.2), (6.3), (6.8), (6.9), and (6.10), it can be observed that a generalized washout filter-based power sharing strategy are formed by the band-pass filter (BPF), realized by cascading LPF and HPF. Therefore, the parameters of washout-based control method are mainly affected by the cut-off frequencies of BPF. The frequency characteristics of a BPF in the washout filter-based controlled islanded MG can be shown in Fig. 6.2, where f_{high} and f_{low} represent the cut-off frequency of the high-pass and low-pass filter, respectively, and f_{center} is the center frequency of the BPF.

The cut-off frequency is defined as the frequency when the power of a signal is at its −3 dB attenuation point. In this way, a good performance of secondary control can be ensured, when the cut-off frequency of the high-pass filter satisfies the following conditions:

Fig. 6.2 The frequency characteristics of a BPF in the secondary controlled islanded MG

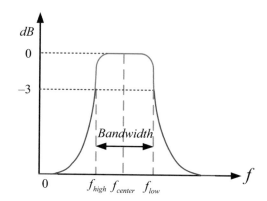

$$\frac{k_{iE}}{k_{pE}+1} < \omega_c, \quad \frac{k_{i\omega}}{k_{p\omega}+1} < \omega_c \qquad (6.11)$$

Equation (6.11) gives a restrictive condition to design the parameters in a washout filter-based or secondary control strategies. Moreover, the physical meaning of parameters of secondary controllers is used to form a BPF, which has not been discussed in the existing literatures. If an islanded MG system with the secondary controllers does not satisfy the conditions imposed by the cut-off frequency restraint as represented by (6.11), the power signals p and q passing through ill-conditioned BPFs will be augmented and oscillating. In other words, droop control may be ineffective, and the dynamic stability of the whole system cannot be guaranteed. Moreover, the stability of islanded MGs with generalized washout filter-based approach and the parameters design guidelines of this equivalent control model will be analyzed in next section.

6.3 Small-Signal Model of Generalized Washout Filter-Based Control Method

This section presents the small-signal model of the generalized washout filter-based power sharing strategy for the islanded microgrid, emphasizing the design of the control parameters, and the stability analysis of the MG.

6.3.1 Power Control Loops

The linearized small-signal models of the active and reactive power controllers can be written as [16]:

$$\begin{cases} \Delta\dot{P} = -\omega_c\Delta P + \omega_c\left(i_{od}\Delta v_{od} + i_{oq}\Delta v_{oq} + v_{od}\Delta i_{od} + v_{oq}\Delta i_{oq}\right) \\ \Delta\dot{Q} = -\omega_c\Delta Q + \omega_c\left(-i_{oq}\Delta v_{od} + i_{od}\Delta v_{oq} + v_{oq}\Delta i_{od} - v_{od}\Delta i_{oq}\right) \end{cases} \quad (6.12)$$

where "Δ" is the small signal perturbation.

By linearizing (6.8) and (6.9), the small-signal dynamics of the generalized washout filter-based control equations can be obtained as:

$$\begin{cases} \Delta\dot{\omega} = -\dfrac{k_{i\omega}}{k_{p\omega}+1}\Delta\omega - \dfrac{m_p}{k_{p\omega}+1}\Delta\dot{P} \\ \Delta\dot{v} = -\dfrac{k_{i\omega}}{k_{p\omega}+1}\Delta v - \dfrac{m_p}{k_{p\omega}+1}\Delta\dot{Q} \end{cases} \quad (6.13)$$

Besides, voltage phase angle and amplitude expressions in the d-q coordinate system are denoted as:

$$\delta = \arctan\left(\dfrac{v_d}{v_q}\right), \quad v = \sqrt{v_d^2 + v_q^2} \quad (6.14)$$

where v_d and v_q are the projection of the output voltage v of power controllers on two perpendicular rotating d and q axes, respectively, and the phase angle δ between v and v_d is represented by the following equation can be obtained:

$$\begin{cases} \Delta\delta = -\dfrac{v_q}{v_d^2 + v_q^2}\Delta v_d + \dfrac{v_d}{v_d^2 + v_q^2}\Delta v_q \\ \Delta v = \dfrac{v_d}{\sqrt{v_d^2 + v_q^2}}\Delta v_d + \dfrac{v_q}{\sqrt{v_d^2 + v_q^2}}\Delta v_q \\ \Delta\dot{v} = \dfrac{v_d}{\sqrt{v_d^2 + v_q^2}}\Delta\dot{v}_d + \dfrac{v_q}{\sqrt{v_d^2 + v_q^2}}\Delta\dot{v}_q \end{cases} \quad (6.15)$$

By using $\Delta\omega(s) = s\Delta\delta(s)$, and combining (6.12), (6.13) and (6.15), the small-signal model of power stage of each DG inverter can be obtained as:

$$\begin{bmatrix} \Delta\dot{\omega} \\ \Delta\dot{P} \\ \Delta\dot{Q} \\ \Delta\dot{v}_d \\ \Delta\dot{v}_q \end{bmatrix} = [\mathbf{T_{BPF}}]_{5\times5}\begin{bmatrix} \Delta\omega \\ \Delta P \\ \Delta Q \\ \Delta v_d \\ \Delta v_q \end{bmatrix} \quad (6.16)$$

where the complete matrix $\mathbf{T_{BPF}}$ is given in Appendix A.

6.3.2 Equations of Voltage, Current Loop Controllers, and LCL Filters

The output reference current and the linearized small-signal state-space form of the voltage controller, where the standard PI controllers are used, are represented by (6.17) and (6.18), respectively [16]:

$$
\begin{bmatrix} \Delta i_{ld}^{*} \\ \Delta i_{lq}^{*} \end{bmatrix} = \begin{bmatrix} \mathbf{C_V} \end{bmatrix} \begin{bmatrix} \Delta \phi_d \\ \Delta \phi_q \end{bmatrix} + \begin{bmatrix} \mathbf{D_{V1}} \end{bmatrix} \begin{bmatrix} \Delta v_d \\ \Delta v_q \end{bmatrix} + \begin{bmatrix} \mathbf{D_{V2}} \end{bmatrix} \begin{bmatrix} \Delta i_{ld} \\ \Delta i_{lq} \\ \Delta v_{od} \\ \Delta v_{oq} \\ \Delta i_{od} \\ \Delta i_{oq} \end{bmatrix}
\tag{6.17}
$$

$$
\begin{bmatrix} \Delta \dot{\phi}_d \\ \Delta \dot{\phi}_q \end{bmatrix} = \begin{bmatrix} \mathbf{0_V} \end{bmatrix} \begin{bmatrix} \Delta \phi_d \\ \Delta \phi_q \end{bmatrix} + \begin{bmatrix} \mathbf{B_{V1}} \end{bmatrix} \begin{bmatrix} \Delta v_d \\ \Delta v_q \end{bmatrix} + \begin{bmatrix} \mathbf{B_{V2}} \end{bmatrix} \begin{bmatrix} \Delta i_{ld} \\ \Delta i_{lq} \\ \Delta v_{od} \\ \Delta v_{oq} \\ \Delta i_{od} \\ \Delta i_{oq} \end{bmatrix}
\tag{6.18}
$$

where the complete matrix $\mathbf{C_V}$, $\mathbf{D_{V1}}$, $\mathbf{D_{V2}}$, and $\mathbf{B_{V2}}$ are given in Appendix A. $\mathbf{B_{V1}}$ is a second-order identity matrix and $\mathbf{0_V}$ is a second-order zero matrix. The state variables $i*_{ld}$, $i*_{lq}$ and i_{ld}, i_{lq} are the dq-axis commanded filter inductor currents and inverter currents, respectively. The state variables v_{od}, v_{oq} and i_{od}, i_{oq} are the dq-axis actual output voltages and currents of inverters, respectively. ϕ_d and ϕ_q are introduced to establish the small-signal model of the voltage controller.

$$
\begin{bmatrix} \Delta v_{id}^{*} \\ \Delta v_{iq}^{*} \end{bmatrix} = \begin{bmatrix} \mathbf{C_C} \end{bmatrix} \begin{bmatrix} \Delta \gamma_d \\ \Delta \gamma_q \end{bmatrix} + \begin{bmatrix} \mathbf{D_{C1}} \end{bmatrix} \begin{bmatrix} \Delta i_{ld}^{*} \\ \Delta i_{lq}^{*} \end{bmatrix} + \begin{bmatrix} \mathbf{D_{C2}} \end{bmatrix} \begin{bmatrix} \Delta i_{ld} \\ \Delta i_{lq} \\ \Delta v_{od} \\ \Delta v_{oq} \\ \Delta i_{od} \\ \Delta i_{oq} \end{bmatrix}
\tag{6.19}
$$

$$\begin{bmatrix} \Delta\dot{\gamma}_d \\ \Delta\dot{\gamma}_q \end{bmatrix} = \begin{bmatrix} \mathbf{0}_C \end{bmatrix} \begin{bmatrix} \Delta\gamma_d \\ \Delta\gamma_q \end{bmatrix} + \begin{bmatrix} \mathbf{B}_{C1} \end{bmatrix} \begin{bmatrix} \Delta i_{ld}^* \\ \Delta i_{lq}^* \end{bmatrix} + \begin{bmatrix} \mathbf{B}_{C2} \end{bmatrix} \begin{bmatrix} \Delta i_{ld} \\ \Delta i_{lq} \\ \Delta v_{od} \\ \Delta v_{oq} \\ \Delta i_{od} \\ \Delta i_{oq} \end{bmatrix} \qquad (6.20)$$

The output reference voltage and the linearized small-signal state-space form of PI current controller are achieved by (6.19) and (6.20), respectively [17], where the complete matrix $\mathbf{C}_C$, $\mathbf{D}_{C1}$, $\mathbf{D}_{C2}$, and $\mathbf{B}_{C2}$ are presented in Appendix A. $\mathbf{B}_{C1}$ is a second-order identity matrix and $\mathbf{0}_C$ is a second-order zero matrix. The state variables $v* \; id$ and $v* \; iq$ are the dq-axis commanded voltages. γ_d and γ_q are used for making convenience to establish the small-signal model of the current controller.

Besides, the small-signal model of the output LCL filter can be represented with the following state Eqs. [6.18] [18]:

$$\begin{bmatrix} \Delta\dot{i}_{ld} \\ \Delta\dot{i}_{lq} \\ \Delta\dot{v}_{od} \\ \Delta\dot{v}_{oq} \\ \Delta\dot{i}_{od} \\ \Delta\dot{i}_{oq} \end{bmatrix} = \begin{bmatrix} \mathbf{A}_{LCL} \end{bmatrix} \begin{bmatrix} \Delta i_{ld} \\ \Delta i_{lq} \\ \Delta v_{od} \\ \Delta v_{oq} \\ \Delta i_{od} \\ \Delta i_{oq} \end{bmatrix} + \begin{bmatrix} \mathbf{B}_{LCL1} \end{bmatrix} \begin{bmatrix} \Delta v_{id} \\ \Delta v_{iq} \end{bmatrix} + \begin{bmatrix} \mathbf{B}_{LCL2} \end{bmatrix} \begin{bmatrix} \Delta v_{bd} \\ \Delta v_{bq} \end{bmatrix} + \begin{bmatrix} \mathbf{B}_{LCL3} \end{bmatrix} \begin{bmatrix} \Delta\omega \end{bmatrix} (6.21)$$

where the complete matrix $\mathbf{A}_{LCL}$, $\mathbf{B}_{LCL1}$, $\mathbf{B}_{LCL2}$, and $\mathbf{B}_{LCL3}$ are given in Appendix A. The state variables $v \; bd$ and $v \; bq$ are the dq-axis voltages at the point of common coupling (PCC).

6.3.3 Equations of Distributed Lines and Loads

The generic RL loads of the MG system are chosen in this chapter, and the state equations of the RL load connected at PCC are depicted by (6.22) as:

$$\begin{bmatrix} \Delta\dot{i}_{loadD} \\ \Delta\dot{i}_{loadQ} \end{bmatrix} = \begin{bmatrix} -\dfrac{r_{load}}{L_{load}} & \omega_0 \\ -\omega_0 & -\dfrac{r_{load}}{L_{load}} \end{bmatrix} \begin{bmatrix} \Delta i_{loadD} \\ \Delta i_{loadQ} \end{bmatrix} + \begin{bmatrix} \dfrac{1}{L_{load}} & 0 \\ 0 & \dfrac{1}{L_{load}} \end{bmatrix} \begin{bmatrix} \Delta b_D \\ \Delta b_Q \end{bmatrix} + \begin{bmatrix} I_{loadQ} \\ -I_{loadD} \end{bmatrix} \begin{bmatrix} \Delta\omega \end{bmatrix} (6.22)$$

where the state variables i_{loadD}, i_{loadQ} are the dq-axis load currents at the PCC. r_{load} and L_{load} are the load resistance and inductance, respectively.

Besides, the resistance and inductance of the distribution line connected between the ith DG unit (DG_i) and the jth DG unit (DG_j) are represented as follows [19–22]:

$$
\begin{bmatrix} \Delta \dot{i}_{lineDij} \\ \Delta \dot{i}_{lineQij} \end{bmatrix} = \begin{bmatrix} -\dfrac{r_{line}}{L_{line}} & \omega_0 \\ -\omega_0 & -\dfrac{r_{line}}{L_{line}} \end{bmatrix} \begin{bmatrix} \Delta i_{lineDij} \\ \Delta i_{lineQij} \end{bmatrix} + \begin{bmatrix} \dfrac{1}{L_{line}} & 0 & -\dfrac{1}{L_{line}} & 0 \\ 0 & \dfrac{1}{L_{line}} & 0 & -\dfrac{1}{L_{line}} \end{bmatrix} \begin{bmatrix} \Delta b_{Di} \\ \Delta b_{Qi} \\ \Delta b_{Dj} \\ \Delta b_{Qj} \end{bmatrix} + \begin{bmatrix} I_{lineQ} \\ -I_{lineD} \end{bmatrix} [\Delta \omega] \quad (6.23)
$$

where the state variables $i_{lineDij}$ and $i_{lineQij}$ are the dq-axis line currents between the ith and jth bus. r_{line} and L_{line} are the resistance and inductance of distribution lines, respectively.

Moreover, the virtual resistor is assumed to be connected at the inverter bus and the following equation can be obtained by using Kirchhoff's Voltage Law (KVL) [20, 21].

$$
\begin{cases} v_{bD} = r_N \left(i_{oD} - i_{loadD} + i_{lineDij} \right) \\ v_{bQ} = r_N \left(i_{oQ} - i_{loadQ} + i_{lineQij} \right) \end{cases} \quad (6.24)
$$

where r_N is the virtual resistor connected at the ith bus, which is used to increase the dynamic stability of the system and make convenience to establish the small-signal model of the system.

6.3.4 Reference Frame Transformation

Note that each DG operates in its own local reference frame. Therefore, the individual reference frame of a DG needs to be taken as a common reference frame and the rest of all DG units including network and loads are transformed to this reference frame as defined in (6.25) and (6.26):

$$
\begin{bmatrix} x_D \\ x_Q \end{bmatrix}_{global} = \begin{bmatrix} \cos\theta & \sin\theta \\ -\sin\theta & \cos\theta \end{bmatrix} \begin{bmatrix} x_d \\ x_q \end{bmatrix}_{local} \quad (6.25)
$$

$$
\begin{bmatrix} x_d \\ x_q \end{bmatrix}_{local} = \begin{bmatrix} \cos\theta & -\sin\theta \\ \sin\theta & \cos\theta \end{bmatrix} \begin{bmatrix} x_D \\ x_Q \end{bmatrix}_{global} \quad (6.26)
$$

where the state variables x_d, x_q and x_D, x_Q are the dq-axis local and global variables, respectively.

6.3.5 Linearized Model of Complete DG System

From the previous theoretical analysis, it can be deduced that each DG contains 17 state variables and each model of line connected between two DG units contains two state variables. A total number of 36 state variables contained in an islanded MG system considering two parallel DG units is taken as an example [21, 22]:

$$\left[\Delta \dot{\mathbf{X}} \right] = \left[\Delta \dot{\mathbf{X}}_1 \ \Delta \dot{\mathbf{X}}_{1,2} \ \Delta \dot{\mathbf{X}}_2 \right]^T = \mathbf{A}_{sys} \left[\Delta \mathbf{X} \right] \tag{6.27}$$

where $\Delta \mathbf{X}_1, \Delta \mathbf{X}_{1,2}$, and $\Delta \mathbf{X}_2$ are represented by (6.28), (6.29) and (6.30), respectively:

$$\left[\Delta \dot{\mathbf{X}}_1 \right] = [\Delta \dot{\omega}_1 \ \Delta \dot{P}_1 \ \Delta \dot{Q}_1 \ \Delta \dot{v}_{d1} \ \Delta \dot{v}_{q1} \ \Delta \dot{\phi}_{d1} \ \Delta \dot{\phi}_{q1} \ \Delta \dot{\gamma}_{d1} \ \Delta \dot{\gamma}_{q1}$$
$$\Delta \dot{i}_{ld1} \ \Delta \dot{i}_{lq1} \ \Delta \dot{v}_{od1} \ \Delta \dot{v}_{oq1} \ \Delta \dot{i}_{od1} \ \Delta \dot{i}_{oq1} \ \Delta \dot{i}_{loadD1} \ \Delta \dot{i}_{loadQ1}]^T \tag{6.28}$$

$$\left[\Delta \dot{\mathbf{X}}_{1,2} \right] = \left[\Delta \dot{i}_{lineD1,2} \ \Delta \dot{i}_{lineQ1,2} \right]^T \tag{6.29}$$

$$\left[\Delta \dot{\mathbf{X}}_2 \right] = [\Delta \dot{\omega}_2 \ \Delta \dot{P}_2 \ \Delta \dot{Q}_2 \ \Delta \dot{v}_{d2} \ \Delta \dot{v}_{q2} \ \Delta \dot{\phi}_{d2} \ \Delta \dot{\phi}_{q2} \ \Delta \dot{\gamma}_{d2} \ \Delta \dot{\gamma}_{q2}$$
$$\Delta \dot{i}_{ld2} \ \Delta \dot{i}_{lq2} \ \Delta \dot{v}_{od2} \ \Delta \dot{v}_{oq2} \ \Delta \dot{i}_{od2} \ \Delta \dot{i}_{oq2} \ \Delta \dot{i}_{loadD2} \ \Delta \dot{i}_{loadQ2}]^T \tag{6.30}$$

Figure 6.3 shows a sparsity pattern of $\mathbf{A}_{sys}$, where seven regions are depicted in the matrix diagram and the nonzero elements are distributed across the diagonal of the matrix. Moreover, regions 1, 2, and 3 are formed by DG_1, while regions 5, 6, and 7 are formed by DG_2. Region 4 is formed by the distribution line between DG_1 and DG_2. Region 1 and 5 are formed based on (6.16), (6.18), and (6.20), which contain power stages, and the voltage and current control loops. Besides, the angular frequency of DG_1 is set as the reference angular frequency for the DG_2. Region 2 and 6 are formed by using (6.21), which contains LCL filters of each DG unit. Region 3 and 7 are formed by the loads, and virtual resistors are depicted in region 4. Therefore, a new sparsity state matrix diagram can be obtained, where additional patterns identical to those in regions 1, 2, 3, and 4 are located on the diagonal of the matrix following the sparsity pattern of DG_2, when other DG units are added to the MG system.

6.3.6 Small-Signal Analysis

A complete model of the test system was established by a sparsity state matrix $\mathbf{A}_{sys}$ presented in (6.27), and the complete eigenvalues of the system can be calculated, by using the initial conditions of the system in Table 6.1.

Figure 6.4 shows the eigenvalues of the MG system distributed in a large range of frequency scale, which can be divided into three different clusters due to the

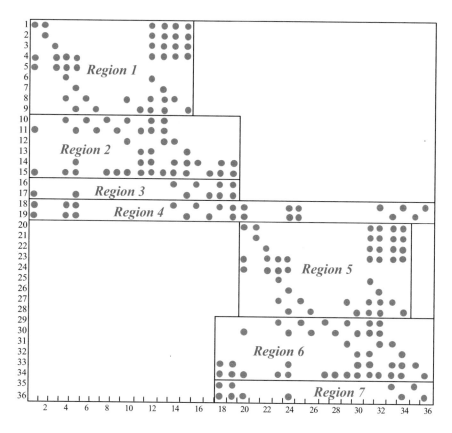

Fig. 6.3 Sparsity pattern of the state matrix $\mathbf{A}_{sys}$

time-scale separation among the different control loops. The cluster "3" appeared in high-frequency modes are sensitive to the distribution of the state variables of LCL filters and the impedance of feeders, while the medium-frequency modes in cluster "2" are affected by the inner control loops. Moreover, it can be observed that the low-frequency modes shown in cluster "1" are sensitive to the state variables (elements in matrix $\mathbf{T_{BPF}}$) of the generalized washout filter-based power controller, which is crucial for analyzing the stability of the three-phase islanded microgrid system.

Figure 6.5 shows a method for the selection of the parameter k_{iE} of the generalized washout filter-based power controllers. The eigenvalue loci of cluster "1" and "2" (the real component is greater than -400) of state matrix $\mathbf{A}_{sys}$ change along with the increasing of ω_{hE} (where $\omega_{hE} = k_{iE}/(k_{pE}+1)$) when the $k_{pE} = 0.001$ and other parameters are chosen as the initial conditions.

From Fig. 6.5, it can be observed that a pair of dominant eigenvalues in cluster 1 and 2 will go across the imaginary axis, which indicates that the system loses stability according to the first Lyapunov's theorem. Therefore, the cut-off frequency ω_{hE}

Table 6.1 Initial conditions and system parameters

Initial conditions and system parameters	Values
LCL filter	$L_f = L_c = 1.8$ mH, $C_f = 25$ μF, $r_{Lf} = 0.1$ Ω and $r_{Lc} = 0.01$ Ω
DC link voltage	650 V
Switching frequency	10 kHz
DG feeder	Feeder 1 inductance and resistance $L_{line1} = 2.2$ mH $r_{line1} = 0.2$ Ω
	Feeder 2 inductance and resistance $L_{line2} = 0.8$ mH $r_{line2} = 0.1$ Ω
Output voltage	DG1: 325.26 V DG2: 325.26 V DG3:325.26 V
PCC voltage	DG1: 322.79 V DG2: 322.79 V DG3:322.79 V
Voltage and current control parameters	**Values**
k_{pv}, k_{iv}	0.175, 200
k_{pc}, k_{ic}	1.8, 131
Power control parameters	**Values**
$k_{p\omega}, k_{i\omega}$	0.005, 4
k_{pE}, ω_c	0.001, 10π
m_p, n_q	10e-4, 10e-4
k_p, k_q	2, 2
Load parameters	**Values**
DG load	Loads inductance and resistance $L_{load1} = L_{load2} = L_{load3} = 720$ mH, $r_{load1} = r_{load2} = r_{load3} = 5$ Ω

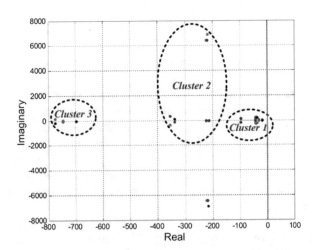

Fig. 6.4 The eigenvalues of the MG system distributed in a large range of frequency scale

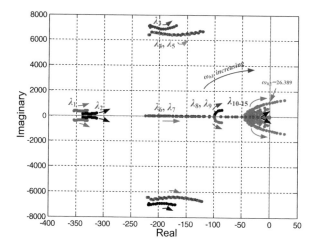

Fig. 6.5 Root locus diagram of parallel DG-based microgrid system with the change of ω_{hE}

of the high-pass filter in (6.10) should be limited and then the corresponding parameter $k_{iE} = 0.6$ can be determined. In addition, other unknown parameters can be easily designed using the similar approach by varying the parameter of controllers
while keeping other parameters fixed.

6.4   Simulation Results Using EMTP

In order to check the difference between the conventional droop control scheme and
the washout filter-based improved droop control scheme, a comparative analysis is
performed by using digital simulation based on EMTP software. The three inverter
units are parallelly connected through feed impedance, and a common resistive load
of 2 Ω is connected, where an abrupt increase of load of another 2 Ω resistance load
is applied from $t = 0.3$ s to $t = 0.7$ s to test the dynamic response of the islanded
microgrid. The rest of the power-stage and control parameters are consistent with
the analysis in the previous section. Both the case of identical and non-identical
power-stage parameters are simulated to show a comparison of the droop characteristics and active and reactive power sharing accuracy.

6.4.1   Conventional Droop Control Scheme

Figure 6.6 shows the simulation results of the conventional droop control scheme
under identical power-stage parameters when droop coefficient $m_p = n_q = 0.0005$.
Figure 6.6a shows the reference voltage amplitude of each inverter. Figure 6.6b

shows the reference frequency of each inverter. Figure 6.6c shows the active and reactive power of each inverter, and it should be noted that the simulation model is implemented in the stationary alpha-beta reference frame. Hence, a ratio of 1.5 should be multiplied when describing the instantaneous three-phase active and reactive power. Besides, Fig. 6.6d shows the tracking error of voltage-loop controller and output voltage in phase "a" of first inverter. In the remainder of this section, the illustration of simulation waveforms is consistent with this case.

Figure 6.7 shows the simulation results of the conventional droop control scheme under identical power-stage parameters when droop coefficient $m_p = n_q = 0.001$. Figure 6.8 shows the simulation results of the conventional droop control scheme under identical power-stage parameters when droop coefficient $m_p = n_q = 0.005$. It can be observed that, when the conventional droop control scheme is applied, the reference voltage amplitude and frequency of each inverter suffer from steady state error, and remarkable transient dip when a sudden increase of load is applied. Besides, when the higher droop coefficient is utilized, higher voltage amplitude and frequency deviations are observed, and significant oscillation would appear with higher voltage tracking error. Hence, it can be concluded that the conventional droop control shows the drawback of voltage amplitude and frequency deviation. Therefore, a smaller droop coefficient could be used to reduce the voltage amplitude and frequency deviation under this scenario.

Figure 6.9 shows the simulation results of the conventional droop control scheme under non-identical power-stage parameters (line inductance of inverter 2 and 3 are +0.1pu and −0.1pu perturbed from the nominal value) when droop coefficient $m_p = n_q = 0.0005$. Figures 6.10 and 6.11 show the identical operating conditions as shown in Fig. 6.9, when the droop coefficient $m_p = n_q = 0.001$, and $m_p = n_q = 0.0011$, respectively.

From Fig. 6.9 to Fig. 6.11, it can be observed that the difference in the line inductance of each inverter results in oscillation in the reference voltage amplitude and frequency, as well as in the instantaneous active power sharing of each inverter. With an increase of the droop coefficient, higher oscillating can be observed in the reference voltage amplitude and frequency, and instantaneous active power with a higher tracking error in the output voltage of each inverter. When the droop coefficient is selected higher than 0.0011, instability would occur under the nominal control parameters. For the practical applications, if the line inductance deviation is high, a reduced droop coefficient would be preferred to ensure the closed-loop stability of the islanded ac microgrid. However, the controller parameter should also be updated using the eigenvalue-based pole-zero placement approach in case of a large line inductance deviation scenario.

Figure 6.12 shows the simulation results of the conventional droop control scheme under non-identical power-stage parameters (both line resistance and inductance of inverter 2 and 3 are +0.1pu and −0.1pu perturbed from the nominal value) when droop coefficient $m_p = n_q = 0.0005$. Figures 6.13 and 6.14 show the same power stage parameters as in Fig. 6.12, when the droop coefficient $m_p = n_q = 0.001$, and $m_p = n_q = 0.0011$, respectively.

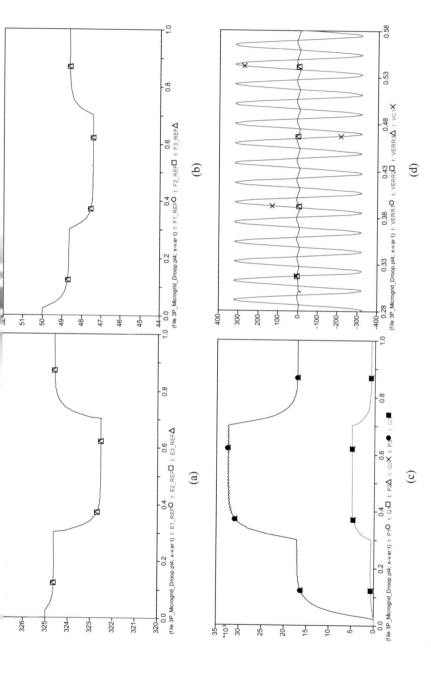

Fig. 6.6 Simulation results of the conventional droop control scheme under identical power-stage parameters when droop coefficient $m_p = n_q = 0.0005$. (**a**) Reference voltage amplitude of each inverter; (**b**) reference frequency of each inverter; (**c**) the active and reactive power of each inverter; (**d**) tracking error of voltage-loop controller and output voltage in phase "a" of first inverter

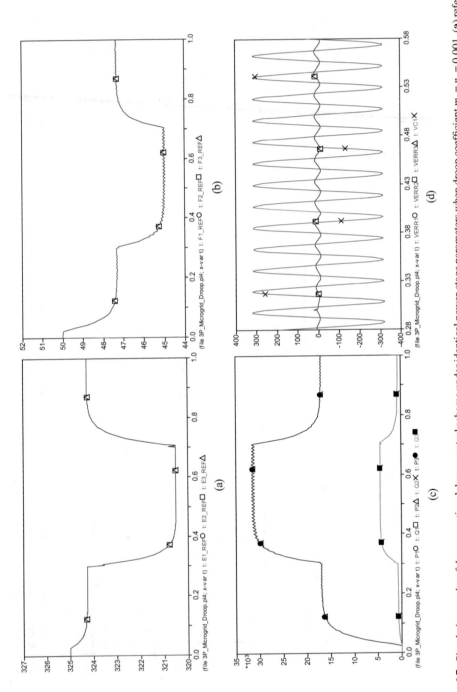

Fig. 6.7 Simulation results of the conventional droop control scheme under identical power-stage parameters when droop coefficient $m_p = n_q = 0.001$. (**a**) reference voltage amplitude of each inverter; (**b**) reference frequency of each inverter; (**c**) the active and reactive power of each inverter; (**d**) tracking error of

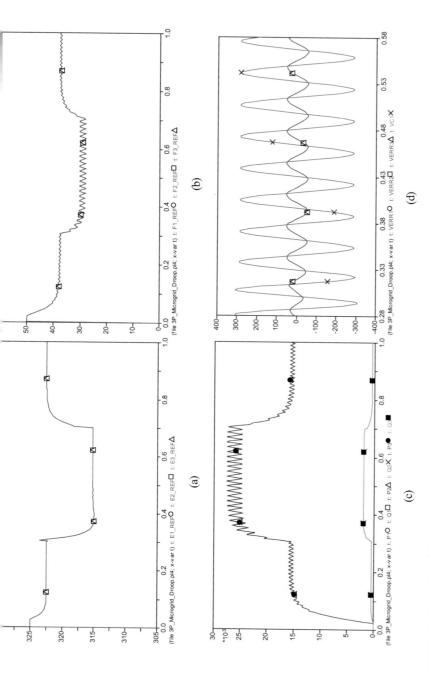

Fig. 6.8 Simulation results of the conventional droop control scheme under identical power-stage parameters when droop coefficient $m_p = n_q = 0.005$. (**a**) Reference voltage amplitude of each inverter; (**b**) reference frequency of each inverter; (**c**) the active and reactive power of each inverter; (**d**) tracking error of voltage-loop controller and output voltage in phase "a" of first inverter

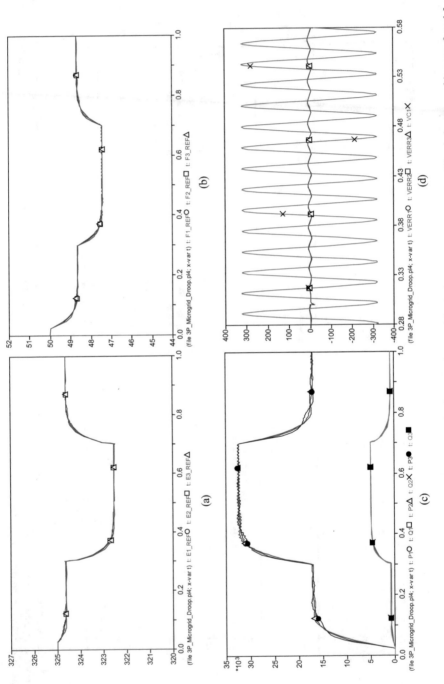

Fig. 6.9 Simulation results of the conventional droop control scheme under non-identical power-stage parameters (line inductance of inverter 2 and 3 are +0.1pu and −0.1pu perturbed from the nominal value) when droop coefficient $m_p = n_q = 0.0005$. (a) reference voltage amplitude of each inverter; (b) reference

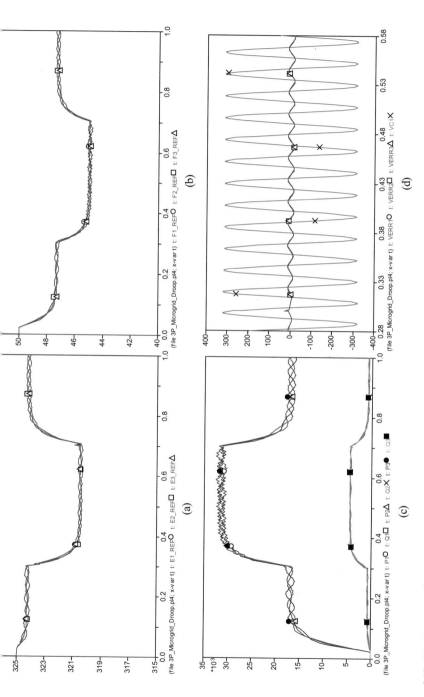

Fig. 6.10 Simulation results of the conventional droop control scheme under non-identical power-stage parameters (line inductance of inverter 2 and 3 are +0.1pu and −0.1pu perturbed from the nominal value) when droop coefficient $m_p = n_q = 0.001$. (**a**) reference voltage amplitude of each inverter; (**b**) reference frequency of each inverter; (**c**) the active and reactive power of each inverter; (**d**) tracking error of voltage-loop controller and output voltage in phase "a" of first inverter

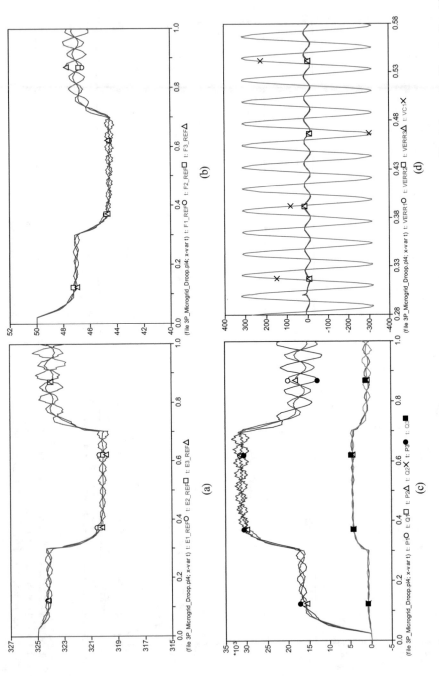

Fig. 6.11 Simulation results of the conventional droop control scheme under non-identical power-stage parameters (line inductance of inverter 2 and 3 are +0.1pu and −0.1pu perturbed from the nominal value) when droop coefficient $m_p = n_q = 0.0011$. (**a**) reference voltage amplitude of each inverter; (**b**) reference f $\ldots$ $\ldots$ (**c**) the active and reactive power of each inverter; (**d**) tracking error of voltage loop controller and output voltage in phase "a" of $\ldots$

From Fig. 6.12 to Fig. 6.14, it can be observed that the difference in the line resistance and inductance of each inverter results in oscillation in the reference voltage amplitude and frequency, as well as in the instantaneous active and reactive power sharing characteristics of each inverter. With an increase of the droop coefficient, higher oscillating deviation can be observed in the reference voltage amplitude and frequency, and instantaneous active and reactive power with a higher tracking error in the output voltage of each inverter. When the droop coefficient is selected higher than 0.0011, instability would occur under the nominal control parameters. Similarly, if the line resistance or inductance deviation is high, a reduced droop coefficient would be preferred to ensure the closed-loop stability of the islanded ac microgrid. However, the controller parameter should also be updated using the eigenvalue-based pole-zero placement approach in case of a high line resistance and inductance deviations.

6.4.2   Washout Filter-Based Improved Droop Control Scheme

Figure 6.15 shows the simulation results of the washout filter-based droop control scheme under identical power-stage parameters when droop coefficient $m_p = n_q = 0.0005$. Figures 6.16 and 6.17 show the same power stage parameters as in Fig. 6.15, when the droop coefficient $m_p = n_q = 0.001$, and $m_p = n_q = 0.005$, respectively.

It can be observed from Figs. 6.15, 6.16, and 6.17 that the accurate power sharing is achieved among the individual inverters, with excellent voltage amplitude and frequency restoration under the transient load perturbations. The voltage amplitude and frequency reference signals recover to the nominal value in a few cycles, which shows the benefit compared to the case when the conventional droop control scheme is implemented. With the increase of droop coefficients, there would be slight transient oscillation in the voltage amplitude and frequency reference signals. Similar to the conventional droop control, the droop coefficients might be reduced to reduce the transient dips and remain closed-loop stability.

Figure 6.18 shows the simulation results of the washout filter based improved droop control scheme under non-identical power-stage parameters (line inductances of inverter 2 and 3 are +0.1pu and −0.1pu perturbed from the nominal value) when droop coefficient $m_p = n_q = 0.0005$. Figures 6.19 and 6.20 show the similar case with identical operating conditions as in Fig. 6.18, when the droop coefficient $m_p = n_q = 0.001$, and $m_p = n_q = 0.002$, respectively.

It can be observed from Figs. 6.18 to 6.20 that a compromised power sharing accuracy is achieved among the individual inverters, with excellent voltage amplitude and frequency restoration under the transient load perturbations. The voltage amplitude and frequency reference signals recover to the nominal value in a few cycles, which is unaffected by the line inductance variations. With the increase of droop coefficients, there would be slight transient oscillation in the voltage amplitude and frequency reference signals. The increase of the droop coefficients

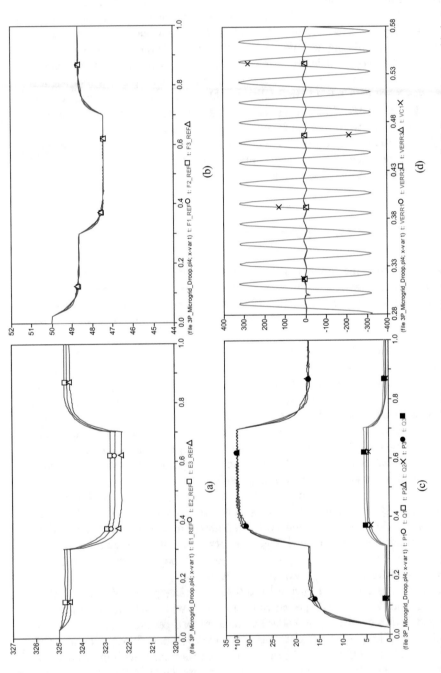

Fig. 6.12 Simulation results of the conventional droop control scheme under non-identical power-stage parameters (both line resistance and inductance of inverter 2 and 3 are +0.1pu and −0.1pu perturbed from the nominal value) when droop coefficient $m_p = n_q = 0.0005$. (**a**) reference voltage amplitude of each

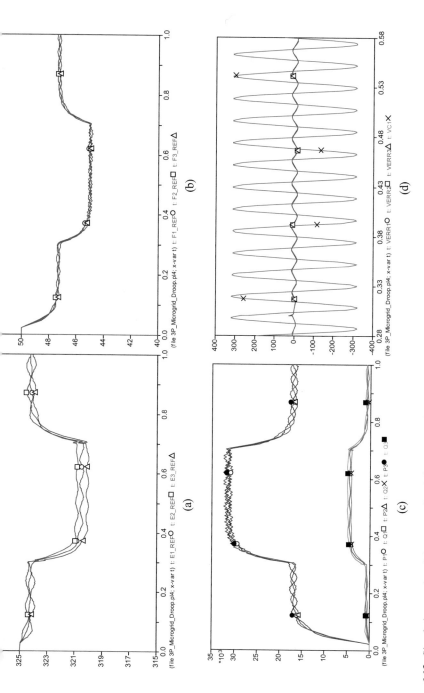

Fig. 6.13 Simulation results of the conventional droop control scheme under non-identical power-stage parameters (both line resistance and inductance of inverter 2 and 3 are +0.1pu and −0.1pu perturbed from the nominal value) when droop coefficient $m_p = n_q = 0.001$. (**a**) reference voltage amplitude of each inverter; (**b**) reference frequency of each inverter; (**c**) the active and reactive power of each inverter; (**d**) tracking error of voltage-loop controller and output voltage in phase "a" of first inverter

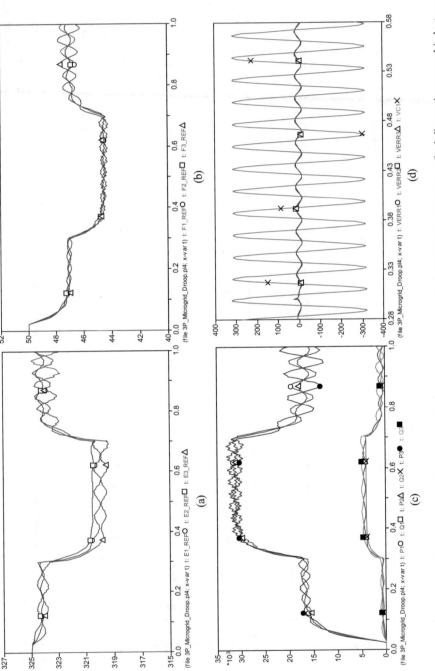

Fig. 6.14 Simulation results of the conventional droop control scheme under non-identical power-stage parameters (both line resistance and inductance of inverter 2 and 3 are +0.1pu and −0.1pu perturbed from the nominal value) when droop coefficient $m_p = n_q = 0.0011$. (**a**) reference voltage amplitude of each inverter. (**b**) reference voltage amplitude of each inverter. (**c**) active and reactive power of each inverter. (**d**) tracking error of voltage loop controller and output

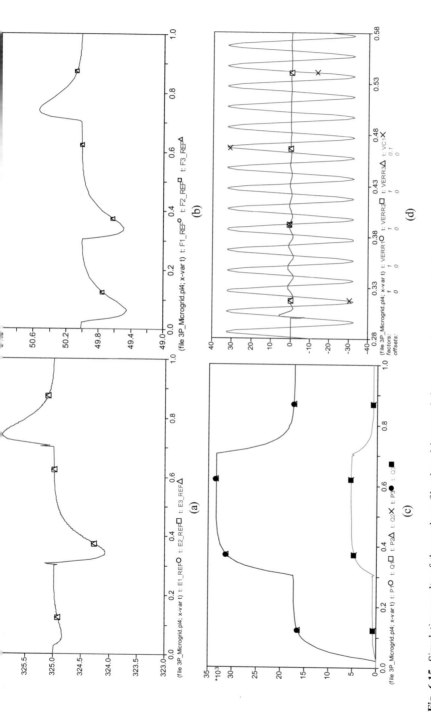

Fig. 6.15 Simulation results of the washout filter-based improved droop control scheme under identical power-stage parameters when droop coefficient $m_p = n_q = 0.0005$. (**a**) reference voltage amplitude of each inverter; (**b**) reference frequency of each inverter; (**c**) the active and reactive power of each inverter; (**d**) tracking error of voltage-loop controller and output voltage in phase "a" of first inverter

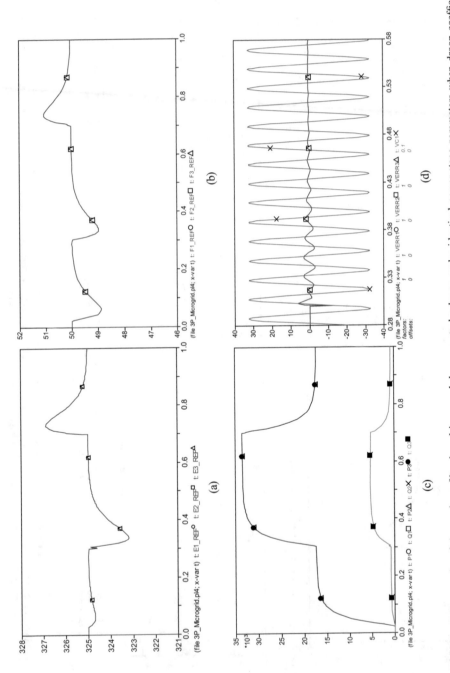

Fig. 6.16 Simulation results of the washout filter-based improved droop control scheme under identical power-stage parameters when droop coefficient $m_n = n_n = 0.001$. (**a**) reference voltage amplitude of each inverter; (**b**) reference frequency of each inverter; (**c**) the active and reactive power of each inverter;

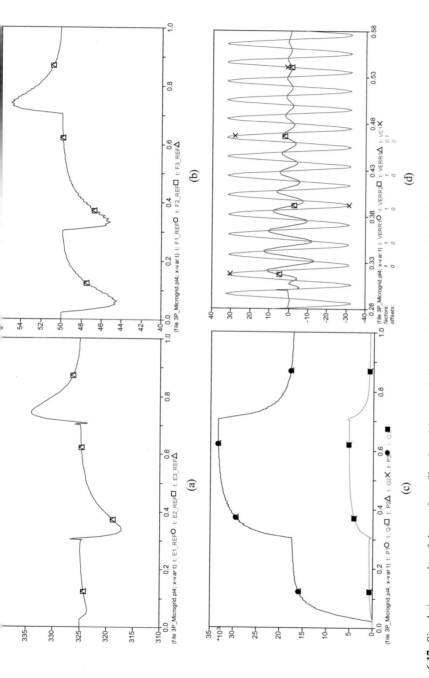

Fig. 6.17 Simulation results of the washout filter-based improved droop control scheme under identical power-stage parameters when droop coefficient $m_p = n_q = 0.005$. (**a**) reference voltage amplitude of each inverter; (**b**) reference frequency of each inverter; (**c**) the active and reactive power of each inverter; (**d**) tracking error of voltage-loop controller and output voltage in phase "a" of first inverter

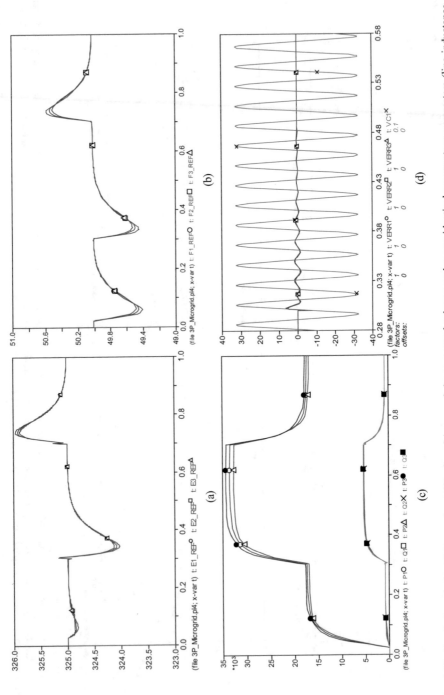

Fig. 6.18 Simulation results of the washout filter-based improved droop control scheme under non-identical power-stage parameters (line inductance of inverter 2 and 3 are +0.1pu and −0.1pu perturbed from the nominal value) when droop coefficient $m_p = n_q = 0.0005$. (**a**) reference voltage amplitude of each inverter. (**b**) reference frequency of each inverter. (**c**) the active and reactive power of each inverter. (**d**) tracking error of voltage-loop controller and output

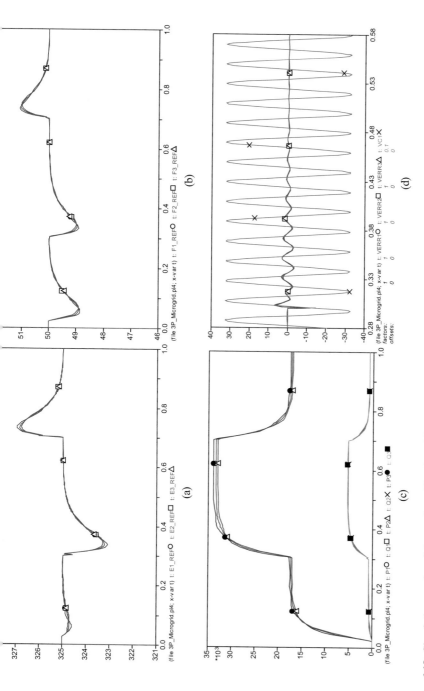

Fig. 6.19 Simulation results of the washout filter-based improved droop control scheme under non-identical power-stage parameters (line inductance of inverter 2 and 3 are +0.1pu and −0.1pu perturbed from the nominal value) when droop coefficient $m_p = n_q = 0.001$. (**a**) reference voltage amplitude of each inverter; (**b**) reference frequency of each inverter; (**c**) the active and reactive power of each inverter; (**d**) tracking error of voltage-loop controller and output voltage in phase "a" of first inverter

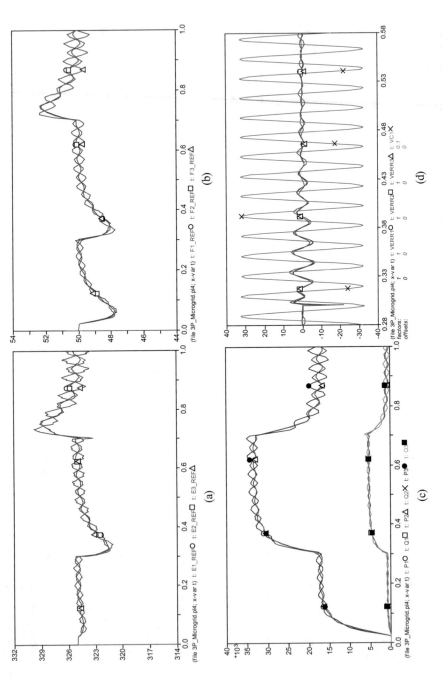

Fig. 6.20 Simulation results of the washout filter-based improved droop control scheme under non-identical power-stage parameters (line inductance of inverter 2 and 3 are +0.1pu and −0.1pu perturbed from the nominal value) when droop coefficient $m_p = n_q = 0.002$. (**a**) reference voltage amplitude of each

improves the accuracy of active power sharing control, with an increased dip in the voltage amplitude and frequency in the inverter output voltages. Whereas, excessive high droop coefficients would cause a higher distortion in the voltage amplitude and frequency reference signals, with a higher tracking error in the voltage tracking control loop.

Figure 6.21 shows the simulation results of the washout filter based improved droop control scheme under non-identical power-stage parameters (both line resistances and inductances of inverter 2 and 3 are +0.1pu and −0.1pu perturbed from the nominal value) when droop coefficient $m_p = n_q = 0.0005$. Figures 6.22 and 6.23 show the similar case with identical operating conditions as in Fig. 6.21, when the droop coefficient $m_p = n_q = 0.001$ and $m_p = n_q = 0.002$, respectively.

It can be observed from Figs. 6.21, 6.22, and 6.23 that a compromised power sharing accuracy is achieved among the individual inverters, with excellent voltage amplitude and frequency restoration under the transient load perturbations. The voltage amplitude and frequency reference signals recover to the nominal value in a few cycles, which is unaffected by the line inductance variations. Similar to the previous case, with the increase of droop coefficients, there would be slight transient oscillation in the voltage amplitude and frequency reference signals. The increase of the droop coefficients improves the accuracy of active power sharing control, with an increased dip in the voltage amplitude and frequency in the inverter output voltages. Similarly, the excessive high droop coefficients would cause a higher distortion in the voltage amplitude and frequency reference signals, with a higher tracking error in the voltage tracking control loop.

6.5   Hardware-in-the-Loop Results

As a supplement analysis to the previous section, the conventional droop control, secondary control considering LBC delay, the washout filter-based strategy and generalized washout filter-based power sharing scheme are implemented on a generic islanded microgrid consists of three parallel DG units, as shown in Fig. 6.24, in order to confirm equivalence between secondary control and washout filter-based control strategies.

In this section, the islanded ac microgrid operates on the unequal feeder impedance and resistive-inductive load conditions. Besides, the different load and feeder impedance conditions are controlled by the switch (SW) 1, 2, 3, and 4. Each DG unit is connected to an LCL filter to eliminate the PWM switching harmonics, and disturbances of DG units, load and feeder impedances are tested to investigate the performance of the active power sharing, frequency, and voltage regulation of the different control strategies. The system parameters are consistent with the theoretical analysis. The conventional droop controller and secondary control are compared with the generalized washout filter-based control scheme, which are implemented in Matlab/Simulink, with measurements recorded through a dSPACE 1006 based real-time digital simulator.

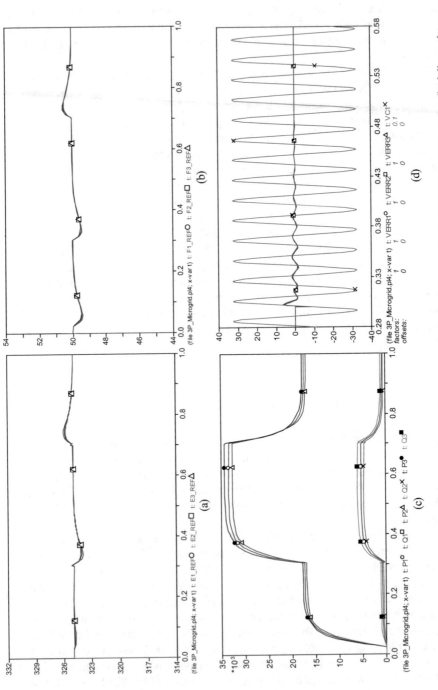

Fig. 6.21 Simulation results of the washout filter-based improved droop control scheme under non-identical power-stage parameters (both line resistance and inductance of inverter 2 and 3 are +0.1 pu and −0.1 pu perturbed from the nominal value) when droop coefficient $m_p = n_q = 0.0005$. (a) Reference voltage amplitude of each inverter. (d) tracking error of each inverter. (d) tracking error of voltage-loop controller

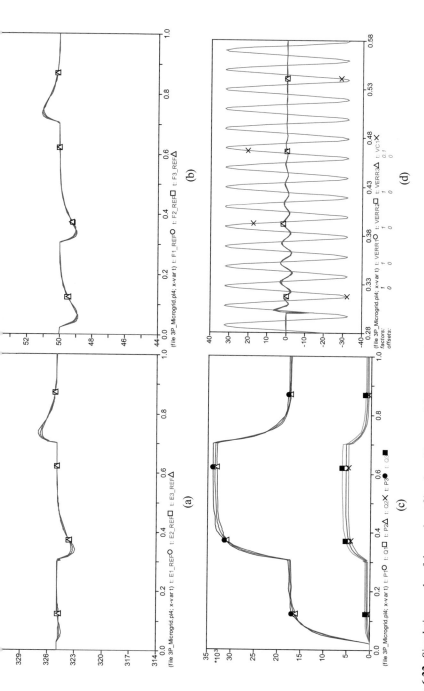

Fig. 6.22 Simulation results of the washout filter-based improved droop control scheme under non-identical power-stage parameters (both line resistance and inductance of inverter 2 and 3 are +0.1pu and −0.1pu perturbed from the nominal value) when droop coefficient $m_p = n_q = 0.001$. (**a**) Reference voltage amplitude of each inverter; (**b**) reference frequency of each inverter; (**c**) the active and reactive power of each inverter; (**d**) tracking error of voltage-loop controller and output voltage in phase "a" of first inverter

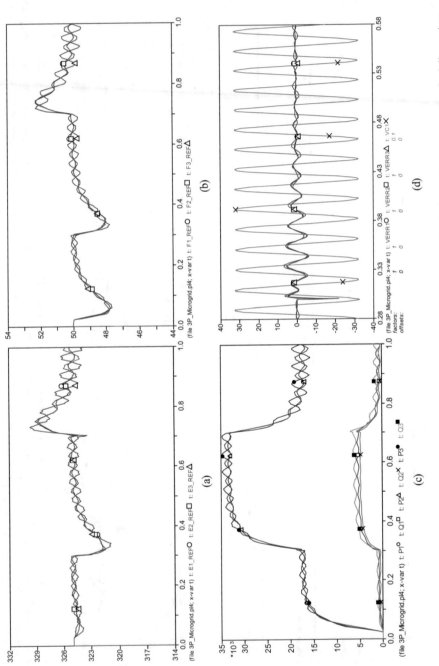

Fig. 6.23 Simulation results of the washout filter-based improved droop control scheme under non-identical power-stage parameters (both line resistance and inductance of inverter 2 and 3 are +0.1pu and −0.1pu perturbed from the nominal value) when droop coefficient $m_p = n_q = 0.002$. (a) Reference voltage amplitude, and reactive power of each inverter. (d) tracking error of voltage loop controller

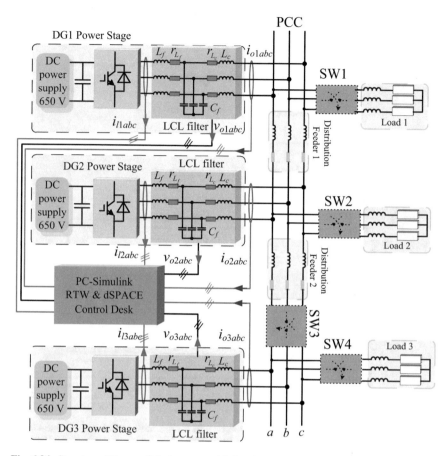

Fig. 6.24 Structure of the paralleled-connected DG units in an islanded AC MG

6.5.1   Performance of Conventional Droop Controller

In Fig. 6.25, the active power, frequency, and voltage amplitude from each inverter operating under a conventional droop control scheme are shown. Initially, the microgrid operates in the steady-state under no load condition. At $t = 1$ s, the load 1 is connected to the microgrid, and the droop mechanism ensures that the active power is shared among the inverters. However, the steady-state errors about 0.14 Hz in the frequency and 0.11 V in the voltage amplitude can be observed. At $t = 2$ s, all loads are connected to the microgrid, and the larger voltage and frequency deviations about 0.18 V and 0.24 Hz, respectively, are caused by the droop control.

Another drawback of the conventional droop control is that the dynamic stability of active power is poor, as shown in Fig. 6.26. In this scenario, the SW3 is disconnected at $t = 1$ s and reconnected at $t = 3$ s, and only DG$_3$ supply the energy to the load "3" during this time. When SW3 is reconnected, it can be observed that the

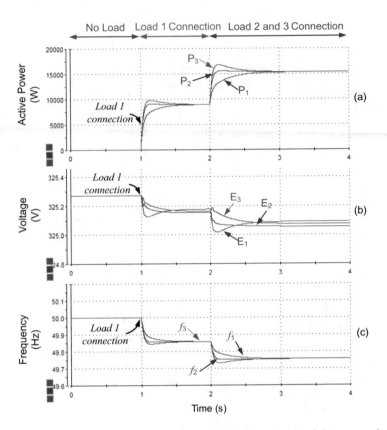

Fig. 6.25 Dynamic response of the islanded microgrid for the conventional droop control under load disturbance conditions. (**a**) active power of each DG unit. (**b**) output voltage of each DG unit. (**c**) frequency of the microgrid

active power, frequency and voltage amplitude differences of DG$_3$ reach to 200%, 1%, and 0.6% of nominal value, respectively. Therefore, the conventional droop control should be further improved to achieve an accurate and robust active power sharing for MGs.

6.5.2   Performance of the Secondary Control Considering the LBC Delays and Communication Failure

The performance of the secondary control strategy applied to a microgrid has been depicted in Figs. 6.27 and 6.28. As seen in Figs. 6.27 and 6.28, the response time and LBC delay are represented by τ and τ_d, respectively. In secondary controlled microgrid, the LBC lines are utilized to send secondary control signals to the

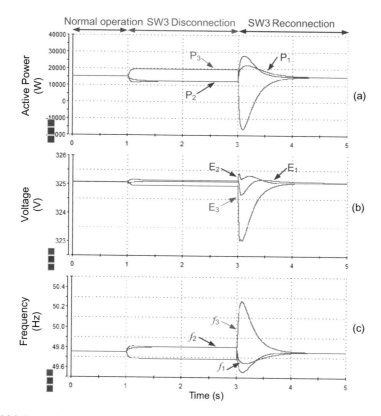

Fig. 6.26 Dynamic response of the islanded microgrid for the conventional droop control under feeder disturbance conditions. (**a**) active power of each DG unit. (**b**) output voltage of each DG unit. (**c**) frequency of the microgrid

primary control level of each DG, in order to restore the frequency and voltage amplitude to the rated values.

In Fig. 6.27, the secondary control is activated at $t = 1$ s, and the voltage and frequency deviation need a time delay to be eliminated, where the LBC delay $\tau_d = 120$ ms is considered. To create a realistic failure scenario, at $t = 3$ s, load 2 and 3 are connected but the LBC lines are deactivated. It is undesirable that the frequencies of each DG unit drop for 0.1 Hz in the steady-state. Moreover, the voltage differences of the DG_1, DG_2, and DG_3 drop for 0.135, 0.149, and 0.175 V in the steady-state, respectively.

Figure 6.28 shows a scenario that the SW3 is connected at $t = 1$ s and disconnected at $t = 3$ s. Besides, the secondary control is activated at $t = 0.5$ s and $\tau_d = 120$ ms is considered in this situation. It can be observed that the active power, frequency and voltage amplitude fluctuation will occur, because the feeder disturbance and LBC delays exist simultaneously in the MG system.

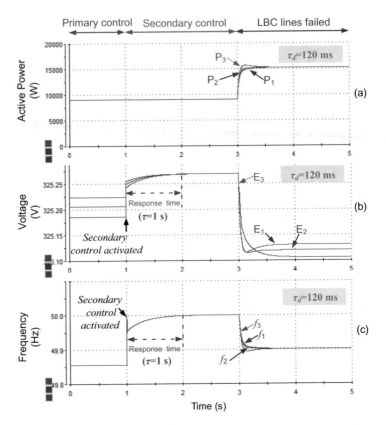

Fig. 6.27 Dynamic response of the islanded microgrid for the secondary control considering LBC delay. (**a**) active power of each DG unit. (**b**) output voltage of each DG unit. (**c**) frequency of the microgrid

6.5.3 Performance of the Washout Filter-Based Control Method

The dynamic response of the washout filter-based control for islanded microgrid system are shown in Figs. 6.29 and 6.30, where the conditions $k_p = 2$, $k_q = 2$ are satisfied. Initially, the microgrid operates in the steady-state under no load condition and the load 1 is connected to the MG system at $t = 1$ s, as shown in Fig. 6.29. Although transient errors about 0.524 Hz in the frequency and about 0.613 V in the voltage amplitude can be obtained, the washout filter-based control strategy is able to eliminate the voltage and frequency deviations in about 1.59 s. At $t = 3$ s, when the load 2 and 3 are connected to the MG system, the frequency and voltage can also be restored to the rated values within 1.5 s.

Figure 6.30 shows the evaluation of dynamic stability of the islanded microgrid system under disturbance of feeders. In this scenario, the SW3 is disconnected at

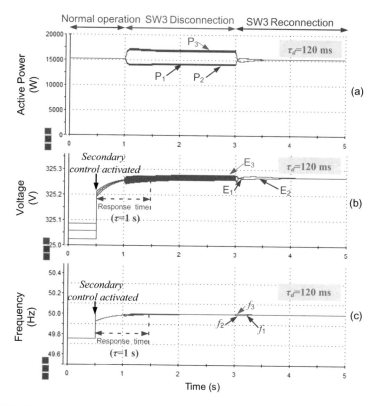

Fig. 6.28 Dynamic response of the islanded microgrid for the secondary control considering feeder disturbances and LBC delay. (**a**) active power of each DG unit. (**b**) output voltage of each DG unit. (**c**) frequency of the microgrid

$t = 1$ s and reconnected at $t = 3$ s. It can be observed that the frequency and voltage can be recovered to the rated values within 1.25 s. However, the voltage and frequency differences of the DG$_2$ reach to 0.72 V and 0.57 Hz, respectively, which are even larger than the deviations caused by the droop control. Therefore, the dynamic stability of the microgrid needs to be further improved, when simultaneously restoring the frequency and voltage amplitude while sharing active power.

6.5.4   Performance of the Generalized Washout Filter-Based Control Strategy

When the generalized washout filter-based control scheme is activated for each DG unit, the precise active power sharing can be achieved in an islanded microgrid, as shown in Fig. 6.31a. Moreover, the effect of unequal load impedances is considered, where inductance and resistance of load 1 are changed to 800 mH and 7 Ω,

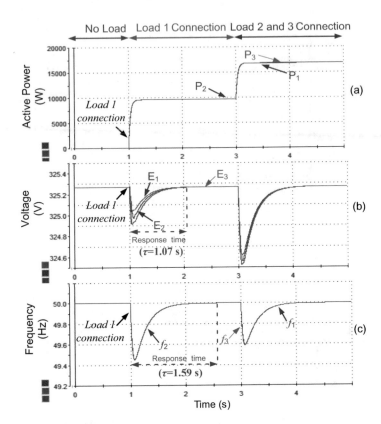

Fig. 6.29 Dynamic response of the islanded microgrid for the washout filter power sharing strategy under load disturbance conditions. (**a**) active power of each DG unit. (**b**) output voltage of each DG unit. (**c**) frequency of the microgrid

respectively. The load 1 is connected at $t = 1$ s and the rest of loads are connected to the microgrid at $t = 2$ s. Compared with the conventional droop control, the dynamic stability of the active power is significantly enhanced in Fig. 6.31a. The same as the secondary control, frequency, and voltage amplitude can be restored to the rated values in a short time (less than 0.68 s). Moreover, compared with the washout filter-based and secondary control methods, there is only a small fluctuation in frequency and voltage amplitude less than 0.8% and 0.012%, respectively, with the disturbance of load impedance as shown in Fig. 6.29.

The effects of feeder and DG disturbances are shown in Fig. 6.32, where the SW3 is disconnected at $t = 1$ s and reconnected at $t = 3$ s. Moreover, unequal load impedances, where inductance and resistance of load 1 are changed to 800 mH and 7 Ω, respectively, are also considered in this scenario. It can be seen that the performance of the active power sharing, frequency, and voltage regulation of the MG system can be ensured, and the difference in the transient behavior is negligible in comparison to the cases in the conventional droop control depicted in Fig. 6.26 and

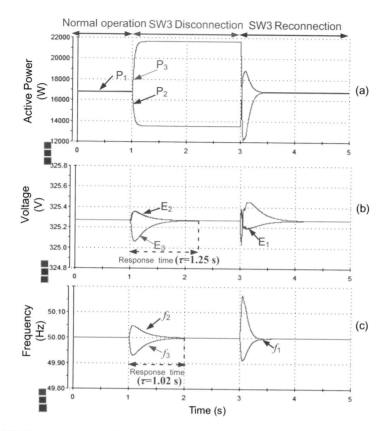

Fig. 6.30 Dynamic response of the islanded microgrid for the washout filter-based control under feeder disturbance conditions. (**a**) active power of each DG unit. (**b**) output voltage of each DG unit. (**c**) frequency of the microgrid

secondary control shown in Fig. 6.28. In addition, compared with the washout filter-based and secondary control methods, the maximum fluctuation of frequency and voltage amplitude less than 0.5% and 0.015%, respectively, with the disturbance of load impedance as shown in Fig. 6.32.

Notably, no communication line is needed in the generalized washout filter-based control scheme, which is immune to the communication delay and data dropout. Although both the washout filter-based control and secondary control can restore the frequency and voltage amplitude to the rated values when sharing the active powers, the generalized washout filter-based power sharing strategy is more robust to the LBC delay, the load/feeder/DG disturbances, and parameter uncertainties. Moreover, the dynamic stability is improved and the fluctuation of frequency and voltage amplitude are decreased significantly, compared with the washout filter-based control method.

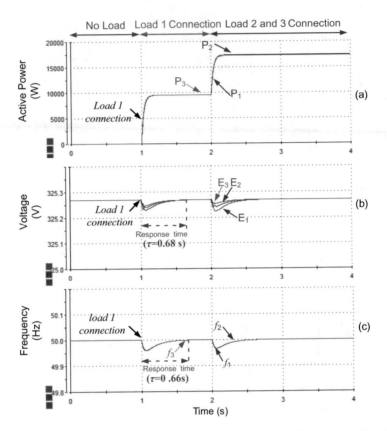

Fig. 6.31 Dynamic response of the islanded microgrid for the generalized washout filter-based control under load disturbance conditions. (**a**) active power of each DG unit. (**b**) output voltage of each DG unit. (**c**) frequency of the microgrid

6.6 Experimental Results

Figure 6.33 shows the experiments on a down-scaled parallel-connected three-phase inverters-based islanded microgrid, in order to evaluate the effectiveness of the generalized washout filter-based power sharing strategy. In addition, the dc-link voltages for each inverter are set as 15 V, with 10 Ω load connected through LCL filters, where $L_{f1} = L_{f2} = 4$ mH and $L_{c1} = L_{c2} = 1.35$ mH, and the capacitor $C_{f1} = C_{f2} = 2.5$ μF. The experimental setup is controlled by a TMS320F28335 digital signal processor (DSP), and controller parameters of the islanded MG system are consistent with the theoretical analysis.

Figure 6.34 shows the experimental results under steady state operation of the generalized washout filter-based control scheme. As shown in Fig. 6.34, output currents of DG_1 and DG_2 in phase "a" are represented by i_{o1a} and i_{o2a}, respectively, and the load current and voltage in phase "a" are represented by i_{loada}, v_{loada}, respectively.

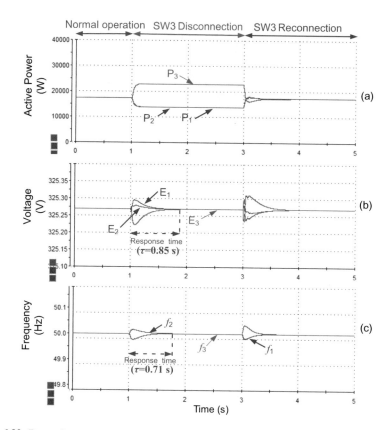

Fig. 6.32 Dynamic response of the islanded microgrid for the generalized washout filter-based control under feeder disturbance conditions. (**a**) active power of each DG unit. (**b**) output voltage of each DG unit. (**c**) frequency of the microgrid

Initially, the DG units are disconnected to the microgrid and currents and voltage are equal to zero. When DG units are abruptly connected to the system, the occurrence of current overshoots can be prevented by the effective voltage and current controllers and power sharing strategy. Moreover, in the steady-state conditions, experimental results indicate that output currents are in phase with the output voltages of the two inverters, and both parallel inverters share current equally. This suggests that the active power sharing is realized by generalized washout filter-based control strategy.

The load voltage THD with the generalized washout filter-based control is less than 5%, which is shown in Fig. 6.35. The negligible output harmonic contents of voltage further validate the effective performance of the current controllers and power sharing strategy. Figures 6.36 and 6.37 show the steady-state and transient response of the generalized washout filter-based control scheme under the disturbance of loads. As shown in Fig. 6.36, the voltage deviations are inevitable when the loads are disconnected to the microgrid system. However, the voltage can be

Fig. 6.33 Experimental setup for the down-scaled prototype AC microgrid

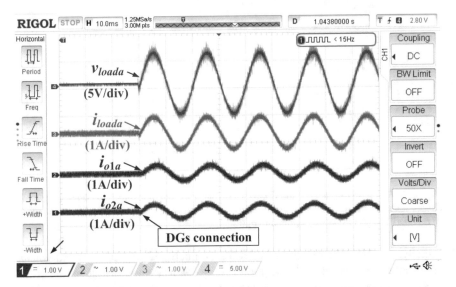

Fig. 6.34 Experimental waveforms of output currents of inverters, and output voltage and current of loads when DG units are connected to the microgrid

restored to the rated values in a short time. Moreover, as depicted in Fig. 6.37, the generalized washout filter-based control strategy can also eliminate the steady-state voltage deviations when the loads are reconnected. To conclude, the experimental results of the dynamic response of the voltage and current further verifies the effectiveness of the generalized washout filter-based control method.

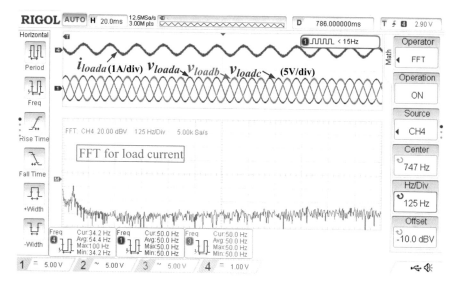

Fig. 6.35 Experimental waveforms of voltage and current of loads, and the FFT-curve of the A-phase load current in islanded microgrid

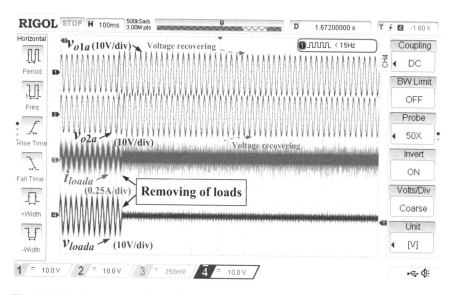

Fig. 6.36 Experimental waveforms of output voltages of inverters, and output voltage and current of loads when loads are disconnected to the microgrid

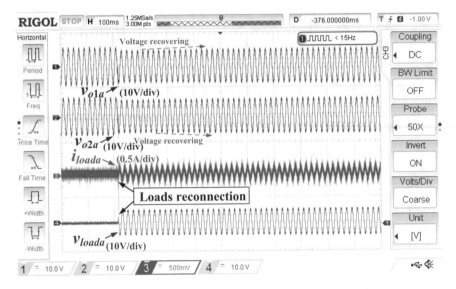

Fig. 6.37 Experimental waveforms of output voltages of inverters, and output voltage and current of loads when loads are reconnected to the microgrid

6.7 Conclusions

This chapter reveals that there is an equivalence between a washout filter-based strategy and secondary control, and the physical meaning of parameters of secondary controllers is discussed, emphasizing that the proportional and integral coefficients of the secondary controller are used to form a band-pass filter. In addition, a generalized washout filter-based power sharing strategy has been obtained to significantly improve the dynamic stability of the system, which can be considered as an enhanced washout filter-based control method. Compared to the secondary control, the generalized washout filter-based control method can eliminate the steady-state errors in the output voltage amplitude and frequency due to the droop control without using LBC links. Compared to the existing washout filter-based control method, the generalized washout filter-based control scheme shows the benefits of enhanced dynamic response under load and feeder disturbances, and reduced overshoots in the output voltages under dynamic disturbances.

A complete small-signal model of the generalized washout filter control scheme is proposed, which can be applied to design of control parameters of the equivalent control model and analyze the stability of the MG. The EMTP-based digital simulation results are given to address the difference between the conventional droop control scheme and the washout filter-based improved droop control approach, and parameter sensitivity of the line impedance is also compared under different droop coefficients conditions. Furthermore, the hardware-in-the-loop (HIL) results of the conventional droop control, secondary control with LBC delay, washout filter-based method, and generalized washout filter-based power sharing scheme are given under unequal feeder impedances and load/DG disturbance conditions to show the effectiveness of the theoretical findings. Finally, the experimental results further validate that the proposed approach are capable to restore the voltages to the rated values without any LBC line and extra control loop, which is more robust to the low bandwidth communication delay and load/DG disturbance.

Appendix A

The matrix $\mathbf{T}_{BPF}$ of power stage is derived as (6.A.1), and the elements in matrix $\mathbf{T}_{BPF}$ are depicted as (6.A.2) and (6.A.3):

$$\mathbf{T}_{BPF} = \begin{bmatrix} -\omega_h & k_2\omega_c & 0 & 0 & 0 \\ 0 & -\omega_c & 0 & 0 & 0 \\ 0 & 0 & -\omega_c & 0 & 0 \\ -v_q & 0 & k_3 v_d & -\dfrac{\omega_h v_d^2}{v_d^2+v_q^2} & -\dfrac{\omega_h v_d v_q}{v_d^2+v_q^2} \\ v_d & 0 & k_3 v_q & -\dfrac{\omega_h v_d v_q}{v_d^2+v_q^2} & -\dfrac{\omega_h v_q^2}{v_d^2+v_q^2} \end{bmatrix}_{5\times5} \tag{6.A.1}$$

$$k_1 = \frac{n_q}{1+k_{pE}}, \quad k_2 = \frac{m_p}{1+k_{p\omega}}, \quad k_3 = \frac{k_1\omega_c\sqrt{v_d^2+v_q^2}}{v_d^2+v_q^2}. \tag{6.A.2}$$

$$\omega_{hE} = \frac{k_{iE}}{1+k_{pE}}, \omega_{h\omega} = \frac{k_{i\omega}}{1+k_{p\omega}}. \tag{6.A.3}$$

The matrices $\mathbf{C_V}$, $\mathbf{D_{V1}}$, $\mathbf{D_{V2}}$, and $\mathbf{B_{V2}}$ in the inner voltage controller are derived as:

$$\mathbf{C_V} = \begin{bmatrix} k_{iv} & 0 \\ 0 & k_{iv} \end{bmatrix}_{2\times2}, \quad \mathbf{D_{V2}} = \begin{bmatrix} 0 & 0 & -k_{pv} & -\omega^*C_f & F & 0 \\ 0 & 0 & \omega^*C_f & -k_{pv} & 0 & F \end{bmatrix}_{2\times6} \tag{6.A.4}$$

$$\mathbf{D_{V1}} = \begin{bmatrix} k_{pv} & 0 \\ 0 & k_{pv} \end{bmatrix}_{2\times2}, \quad \mathbf{C_C} = \begin{bmatrix} k_{ic} & 0 \\ 0 & k_{ic} \end{bmatrix}_{2\times2}, \quad \mathbf{B_{V2}} = \begin{bmatrix} 0 & 0 & -1 & 0 & 0 & 0 \\ 0 & 0 & 0 & -1 & 0 & 0 \end{bmatrix}_{2\times6}. \tag{6.A.5}$$

The matrices $\mathbf{C_C}$, $\mathbf{D_{C1}}$, $\mathbf{D_{C2}}$, and $\mathbf{B_{C2}}$ in the inner current controller are derived as:

$$\mathbf{D_{C2}} = \begin{bmatrix} -k_{pc} & -\omega^*L_f & 0 & 0 & 0 & 0 \\ \omega^*L_f & -k_{pc} & 0 & 0 & 0 & 0 \end{bmatrix}_{2\times6}. \tag{6.A.6}$$

$$\mathbf{D_{C1}} = \begin{bmatrix} k_{pc} & 0 \\ 0 & k_{pc} \end{bmatrix}_{2\times2}, \quad \mathbf{B_{C2}} = \begin{bmatrix} -1 & 0 & 0 & 0 & 0 & 0 \\ 0 & -1 & 0 & 0 & 0 & 0 \end{bmatrix}_{2\times6}. \tag{6.A.7}$$

The matrices $\mathbf{A_{LCL}}$, $\mathbf{B_{LCL1}}$, $\mathbf{B_{LCL2}}$, and $\mathbf{B_{LCL3}}$ in (6.20) are derived as (6.A.8) and (6.A.9).

$$\mathbf{A_{LCL}} = \begin{bmatrix} -\dfrac{r_{L_f}}{L_f} & \omega^* & -\dfrac{1}{L_f} & 0 & 0 & 0 \\[2mm] -\omega^* & -\dfrac{r_{L_f}}{L_f} & 0 & -\dfrac{1}{L_f} & 0 & 0 \\[2mm] \dfrac{1}{C_f} & 0 & 0 & \omega^* & -\dfrac{1}{C_f} & 0 \\[2mm] 0 & \dfrac{1}{C_f} & -\omega^* & 0 & 0 & -\dfrac{1}{C_f} \\[2mm] 0 & 0 & \dfrac{1}{L_c} & 0 & -\dfrac{r_{L_c}}{L_c} & \omega^* \\[2mm] 0 & 0 & 0 & \dfrac{1}{L_c} & -\omega^* & -\dfrac{r_{L_c}}{L_c} \end{bmatrix}_{6\times6} \tag{6.A.8}$$

$$\mathbf{B_{LCL1}} = \begin{bmatrix} \dfrac{1}{L_f} & 0 \\[2mm] 0 & \dfrac{1}{L_f} \\[2mm] 0 & 0 \\ 0 & 0 \\ 0 & 0 \\ 0 & 0 \end{bmatrix}_{6\times2}, \quad \mathbf{B_{LCL2}} = \begin{bmatrix} 0 & 0 \\ 0 & 0 \\ 0 & 0 \\ 0 & 0 \\ -\dfrac{1}{L_c} & 0 \\[2mm] 0 & -\dfrac{1}{L_c} \end{bmatrix}_{6\times2}, \quad \mathbf{B_{LCL3}} = \begin{bmatrix} I_{lq} \\ -I_{ld} \\ V_{oq} \\ -V_{od} \\ I_{oq} \\ -I_{od} \end{bmatrix}_{6\times1}. \tag{6.A.9}$$

References

1. Yu, K., Ai, Q., Wang, S. Y., Ni, J. M., & Lv, T. G. (2016). Microgrid system based on small-signal dynamic model. *IEEE Transactions on Smart Grid, 7*(2), 695–705.
2. Mohamed, Y. A. R. I., & Radwan, A. A. (2011). Hierarchical control system for robust microgrid operation and seamless mode transfer in active distribution systems. *IEEE Transactions on Smart Grid, 6*(4), 352–362.

3. Nutkani, I. U., Loh, P. C., Wang, P., & Blaabjerg, F. (2016). Linear decentralized power sharing schemes for economic operation of AC microgrids. *IEEE Transactions on Industrial Electronics, 63*(1), 225–234.
4. T. Wu, Z. Liu, J. J Liu, S. Wang and Z. Y. You, "A unified virtual power decoupling method for droop-controlled parallel inverters in microgrids," IEEE Transactions on Power Electronics, vol. 31, no. 8, pp. 5587–5603, 2016.
5. Ahmadi, S., Bevrani, H., Shokoohi, S., & Hasanii, E. (2015). An improved droop control for simultaneous voltage and frequency regulation in an AC microgrid using fuzzy logic. In *Proc. 23rd Iranian Conf. on Electrical Engineering (ICEE)* (pp. 1486–1491).
6. Chiang, H. C., Jen, K. K., & You, G. H. (2016). Improved droop control method with precise current sharing and voltage regulation. *IET Generation Transmission and Distribution, 9*(4), 789–800.
7. Dıaz, N. L., Wu, D., Dragicevic, T., Vasquez, J. C., & Guerrero, J. M. (2015). Fuzzy droop control loops adjustment for stored energy balance in distributed energy storage system. In *Proc. 9th International Power Electronics and Motion Control Conf. (ECCE)* (pp. 1–5).
8. He, D. W., Shi, D., & Sharma, R. (2014). Consensus-based distributed cooperative control for microgrid voltage regulation and reactive power sharing. *In Proc. IEEE PES Innov. Smart Grid Technol. Conf. Europe (ISGT-Europe)*, Istanbul, Turkey, (pp. 1–6).
9. Bidram, A., Davoudi, A., & Lewis, F. L. (2014). Two-layer distributed cooperative control of multi-inverter microgrids. *In Proc. IEEE 29th Annu. Appl. Power Electron. Conf. Expo. (APEC)*, Fort Worth, TX, USA, (pp. 2364–2371).
10. Yazdanian, M., & Sani, A. M. (2016). Washout filter-based power sharing. *IEEE Transactions on Smart Grid, 7*(2), 967–968.
11. Guerrero, J. M., Chandorkar, M., Lee, T., & Loh, P. C. (2013). Advanced control architectures for intelligent microgrids-part I: Decentralized and hierarchical control. *IEEE Transactions on Industrial Electronics, 60*(4), 1254–1262.
12. Palizban, O., & Kaohaniemi, K. (2015). Hierarchical control structure in microgrids with distributed generation: island and grid-connected mode. *Renewable and Sustainable Energy Reviews, 44*, 797–813.
13. Shafiee, Q., Stefanovic, C., Dragicevic, T., Popovski, P., Vasquez, J. C., & Guerrero, J. M. (2014). Robust networked control scheme for distributed secondary control of islanded microgrids. *IEEE Transactions on Industrial Electronics, 61*(10), 5363–5374.
14. Wang, P. B., Lu, X. N., Yang, X., Wang, W., & Xu, D. G. (2015). An improved distributed secondary control method for DC microgrids with enhanced dynamic current sharing performance. *IEEE Transactions on Power Electronics, 31*(9), 6658–6673.
15. Zhang, H. G., Kim, S., Sunand, Q. Y., & Zhou, J. G. Distributed adaptive virtual impedance controlfor accurate reactive power sharing basedon consensus control in microgrids. *IEEE Transactions on Smart Grid, to be published.* https://doi.org/10.1109/TSG.2015.2506760
16. Yu, K., Ai, Q., Wang, S. Y., Ni, J. M., & Lv, T. G. (2016). Analysis and optimization of droop controller for microgrid system based on small-signal dynamic model. *IEEE Transactions on Smart Grid, 7*(2), 695–705.
17. Pogaku, N., Prodanovic, M., & Green, T. C. (2006). Inverter-based microgrids: Small-signal modelling and testing. In *Proc. 3rd International Power electronics, machines and drives (PEMD) Conf.*
18. Mohamed, Y. A. R. I., & Saadany, E. F. E. (Nov. 2008). Adaptive decentralized droop controller to preserve power sharing stability of paralleled inverters in distributed generation microgrids. *IEEE Transactions on Power Electronics, 23*(6), 2806–2816.
19. Han, Y., Li, H., Xu, L., Zhao, X., & Guerrero, J. M. (2018). Analysis of washout filter-based power sharing strategy – An equivalent secondary controller for islanded microgrid without LBC lines. *IEEE Transactions on Smart Grid, 9*(5), 4061–4076.
20. Rasheduzzaman, M., Mueller, J. A., & Kimball, J. W. (2014). An accurate small-signal model of inverter-dominated islanded microgrids using dq reference frame. *IEEE Journal of Emerging and Selected Topics in Power Electronics, 2*(4), 1070–1080.

21. Pogaku, N., Prodanovic, M., & Green, T. C. (2007). Modeling, analysis and testing of autonomous operation of an inverter-based microgrid. *IEEE Transactions on Power Electronics, 22*(2), 613–625.
22. Olivares, D. E., Sani, A. M., Etemadi, A. H., Cañizares, C. A., Iravani, R., Kazerani, M., Hajimiragha, A. H., Canizares, C. A., Iravani, R., Kazerani, M., Hajimiragha, A. H., Bellmunt, O. G., Saeedifard, M., Behnke, R. P., Estévez, G. A. J., & Hatziargyriou, N. D. (2014). Trends in microgrid control. *IEEE Transactions on Smart Grid, 5*(4), 1905–1919.

Chapter 7
Consensus-Based Enhanced Droop Control Scheme for Accurate Power Sharing and Voltage Restoration in Islanded Microgrids

In order to further enhance the washout filter-based improved droop control scheme, this chapter presents a consensus-based enhanced droop control scheme for islanded microgrid system, which achieves accurate active and reactive power sharing, while maintaining the frequency recovering and keeping the average voltage to the rated values. In the proposed control scheme, only the neighborhood reactive power information needs to be exchanged by using a sparse low-bandwidth communication (LBC) network instead of delivering information of active power, reactive power and frequency by communication links in the existing consensus methods. Compared to the existing consensus-based methods, the transfer data and data latency are significantly reduced and the high reliability of the system can be achieved. Moreover, the accurate active power sharing and frequency recovering can be ensured under disturbances of load and feeder impedance, even in case of communication failures. Finally, the steady-state performance and local exponential stability analysis of the proposed control scheme are also presented, and the simulation and hardware-in-the-loop (HiL) test results are provided to validate the effectiveness of the proposed control scheme.

7.1 Introduction

In recent years, the distributed generation (DG) using renewable energy resources like wind turbines (WTs), photovoltaics (PVs) and micro-turbines (MTs) has received more and more attention. To integrate different types of DG units and distributed energy storage systems effectively in the distribution networks, the concept of microgrid (MG) has been presented [1–9]. Compared with the conventional distributed generation systems, the microgrids can operate either in the grid-connected mode or islanded mode. When a MG operates in the grid-connected mode, the voltage and frequency are affected by the main grid; how to minimize the cost of power generation, optimize charge–discharge cycles of storage units, and simultaneously

© Springer Nature Switzerland AG 2022
Y. Han, *Modeling and Control of Power Electronic Converters for Microgrid Applications*, https://doi.org/10.1007/978-3-030-74513-4_7

maintain a high-power factor at the point of common coupling (PCC) are crucially important. In an islanded MG, the droop control method has been widely used to establish the system frequency and bus voltages, as well as sharing active powers among the DG units simultaneously without using critical communications. However, in these methods, the voltage amplitude and frequency deviations caused by the traditional droop control methods are inevitable, and the reactive power sharing is poor under unequal feeder impedance conditions [10–12].

In order to deal with the problems mentioned above, the methods based on the virtual impedance or the improved virtual impedance have been proposed to share active and reactive powers among the DG units[13–16]. However, the information about MG parameters and feeders are needed, which limits these methods in practical applications. A typical secondary control strategy is often utilized in the islanded microgrids to restore the voltage and frequency to the rated values in order to compensate the limitations of the droop mechanism, while the reactive power sharing problem has not been fully addressed [17]. An improved secondary control strategy with voltage compensation controller has been presented in [18] to share reactive power and restore the voltage to the rated value. In addition, the centralized secondary control methods are presented to share the reactive power under unequal feeder impedance scenarios. Although the voltage and frequency deviations can be eliminated in these methods, the microgrid central controller (MGCC) is required to connect each DG unit, which increases the system complexity and reduces the scalability and reliability[18–21].

Recently, the multi-agent control theory has been widely studied in the MG systems [22–29]. The main objective of the consensus control is to achieve general agreements among all agents in a network, which requires only neighbor-to-neighbor interactions. In order to achieve accurate active and reactive power sharing in a fully distributed way, some consensus-based control strategies have been presented [24–32]. For instance, a dynamic consensus algorithm is presented in [25] to balance the discharge rate of energy storage system (ESS) and regulate current and active power of the system, considering the ESS capacities and state of-charge (SOC). However, the reactive power sharing has not been considered in this method. An improved consensus-based control strategy is presented in [30] to regulate the active and reactive powers in a sparse communication network. However, the shortcoming of this technique is the absence of the droop mechanism as a backup controller, which degrades the controller functionality in case of communication failure. A virtual impedance-based consensus control method is presented in [31] to share the reactive power regardless of the mismatched line impedance conditions. However, the frequency deviation caused by droop control is not considered and the values of virtual impedances are difficult to be obtained. It should be noted that the performance of multiple DGs and stability of the system would be affected due to the lack of the droop mechanism in the existing consensus-based control methods, in case of communication failure [32].

Some improved consensus control methods have been used to investigate the issue of accurate power sharing in a microgrid [33, 34]. However, the neighboring frequency, voltage, active power and reactive power information are required to be transferred in the communication network for islanded AC microgrids. In this scenario, the high-density data would lead to challenges related to data processing and analysis for a large-scale microgrid system. Moreover, once a larger latency exists in the communication networks or in case of data dropout, a whole system would be unstable [35].

In this chapter, a consensus-based enhanced droop control scheme is proposed to regulate voltage, frequency and active and reactive powers in an islanded AC microgrid system. Different from existing consensus-based methods, only the neighborhood reactive power information is required in a sparse LBC network, which reduces the transfer data and data latency are significantly reduced, and the high reliability of the system can be achieved. In addition, the active power sharing, frequency recovering and stability of the system can be ensured even under a total communication link failure scenario, while proportional reactive power sharing can be achieved and the average voltage can be controlled to the rated value.

The remainder of this chapter is organized as follows: The preliminary of the graph theory is briefly discussed, and the proposed consensus-based enhanced droop control strategy and the steady-state analysis are introduced in Sect. 7.2. The local exponential stability analysis in consensus controller is given in Sect. 7.3. In Sect. 7.4, the simulations and hardware-in-the-loop (HiL) results of the islanded AC MG system are provided to verify the effectiveness of the proposed method. Finally, the concluding remarks are given in Sect. 7.5.

7.2   Proposed Consensus-Based Enhanced Droop Control and Steady-State Performance Analysis

7.2.1   Graph Theory

A simple directed graph can be described by $\mathscr{G} = (\mathscr{V}, \mathbf{E}, \mathscr{A})$, where $\mathscr{V}$ is the set of n vertices and edges $\mathbf{E} \subseteq \mathscr{V} \times \mathscr{V}$ is the set of undirected edges connecting them [30–32]. Suppose there are N nodes in the directed graph $\mathscr{G}$, elements of the adjacency matrix $\mathscr{A} = [\alpha_{ij}] \in \mathbf{R}^{N \times N}$ is defined that $\alpha_{ij} = 0$ if an edge between i and j exists, and $\alpha_{ij} = 0$ if $e_{ij} \notin \mathbf{E}$. The degree matrix $\mathscr{D} = \mathrm{diag}(d_i)$ is defined with elements $d_i = \sum_{j \in Ni} \alpha_{ij}$ and the corresponding Laplacian matrix $\mathbf{L}$ can be derived as $\mathbf{L} = \mathscr{D} - \mathscr{A}$ with elements: $\mathbf{L}_{ij} = -\alpha_{ij}$ if $i \neq j$ and $\mathbf{L}_{ii} = \sum_j \neq {}_i\alpha_{ij}$ if $i = j$.

In a MG system, consensus control methods often utilize the important theorems of Algebraic Graph Theory:

Definition 1 *A graph $\mathscr{G}$ is said to have a spanning tree if it contains a root node, where there exists at least one direct path to every other node* [36–38].∎

Lemma 1 L *is a symmetric positive semi-definite matrix. Zero is a simple eigen value of* L *with a right eigen vector* $\underline{1}$ *and all non-zero eigen values have positive* real parts if and only if $\mathcal{G}$ has a directed spanning tree, that is, $L\underline{1}_n = \underline{0}_n$, $\underline{1}^T_n L = \underline{0}^T_n$ [36–38]. ∎

7.2.2   Proposed Consensus-Based Enhanced Droop Control Scheme for Power Sharing

In an islanded microgrid, the droop control method is usually used to share the total demand for active and reactive powers with the active power/frequency (P/f) droop characteristic and the reactive power/voltage droop (Q/V) characteristic, with no need of critical communication [16–19]. For a large/medium system, the fundamental droop equations for each DG can be written as:

$$\omega_i = \omega_i^* - m_i P_i, \ E_i = E_i^* - n_i Q_i \qquad (7.1)$$

where i is the index representing each DG unit, ω_i^* and E_i^* are the rated angular frequency and voltage, respectively; m_i and n_i are frequency and amplitude droop coefficients, respectively. Additionally, P_i and Q_i are the measured average active and reactive power values, respectively.

However, the voltage and frequency deviations caused by the droop control are inevitable and the reactive power cannot be shared in this method. Therefore, a consensus-based enhanced droop control scheme is proposed in this chapter, as depicted in Fig. 7.1, where the inner loop of VSI-based MG, including voltage and current controllers, is established to regulate the output voltage and current while maintaining the system stability. Note that the reference of the voltage control loop is generated by the proposed reactive power controller.

The proposed consensus-based enhanced droop control scheme can be expressed as follows:

$$\begin{cases} \omega_i = \omega_i^* - m_i \cdot \dfrac{s}{s+\omega_h} \cdot \dfrac{\omega_c}{s+\omega_c} \cdot P_i \\[2ex] E_i = E_i^* - \left(k_p + \dfrac{k_i}{s}\right) \displaystyle\sum_{j \in N_i} b\alpha_{ij} \left(\dfrac{Q_j}{\chi_j} - \dfrac{Q_i}{\chi_i}\right) \end{cases} \qquad (7.2)$$

where p_i is instantaneous active power. ω_i and E_i represent the frequency and amplitude of the output voltage. ω_h and ω_c represent the upper and lower cut-off frequencies of the improved frequency controller, respectively. P_i and Q_i are the measured average active and reactive power values, respectively. m_i, k_p, k_i, and b are positive gains. ω_i^* and E_i^* are the rated angular frequency and voltage, respectively. χ_i is a weighted coefficient, considering capacity of DG_i. which can be set to be equal to the derivative of the droop coefficient of DG_i.

As depicted in Fig. 7.1 and (7.2), the improved frequency droop controller can be utilized to achieve accurate active power sharing and maintain the frequency of each

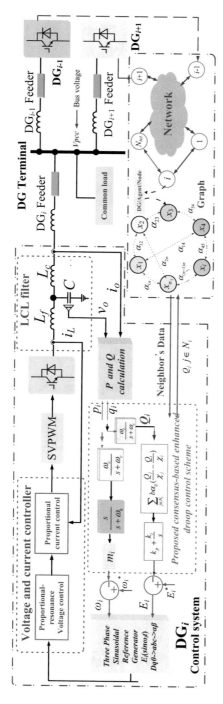

Fig. 7.1 Proposed consensus-based enhanced droop control scheme for DG$_i$ of the islanded microgrid

DG at the rated value without the neighborhood voltage and frequency information. Moreover, the consensus-based reactive power controller can be utilized to share reactive power considering capabilities of DG units, while regulating the average voltage of the MG to the rated value [30–32].

7.2.3 Steady-State Performance Analysis of the Proposed Consensus Control

1. *Reactive Power Sharing*

In the time domain, the reactive power/voltage consensus equation can be rewritten as follows:

$$E_i = E_i^* - \left[k_p \sum_{j \in N_i} b\alpha_{ij} \left(\frac{Q_j}{\chi_j} - \frac{Q_i}{\chi_i} \right) + k_i \int_0^t \sum_{j \in N_i} b\alpha_{ij} \left(\frac{Q_j}{\chi_j} - \frac{Q_i}{\chi_i} \right) d\tau \right] \tag{7.3}$$

By taking the derivative of output voltage of each DG unit with respect to time, the dynamics of reactive power consensus control can be expressed as:

$$\dot{E}_i = -k_i \sum_{j \in N_i} b\alpha_{ij} \left(\frac{Q_j}{\chi_j} - \frac{Q_i}{\chi_i} \right) \tag{7.4}$$

Considering the derivative of output voltage is equal to zero in the steady-state [19], the following matrix can be derived:

$$\dot{E} = -KL\left(Q\chi^{-1} \right) = 0 \tag{7.5}$$

where the Laplacian matrix is represented by L, the diagonal matrix K is defined $K = \mathrm{diag}\{k_1, k_2, \ldots, k_N\}$ and column vectors $\dot{E}$, and Q/χ are expressed as $\dot{E} = \begin{bmatrix} \dot{E}_1 & \dot{E}_2 & \cdots & \dot{E}_N \end{bmatrix}^T$ and $Q\chi^{-1} = [Q1/\chi_1, Q2/\chi_2, \ldots, Q\,N/\chi_N]^T$, respectively.

Recall that only non-zero solution to $L\,X = 0$ is $X = k\underline{1}$ and $k > 0$ and the solution to (7.5) can be obtained as:

$$Q = k(\chi)^{-1} \underline{1} \text{ and } k > 0 \tag{7.6}$$

As shown in (7.6), the proportional reactive power sharing can be realized by the reactive power consensus controller.

2. *Average Voltage Regulation*

Considering $\underline{1}^T_n L = \underline{0}^T_n$, by multiplying $\underline{1}^T_n K^{-1}$ in (7.5), the following equation can be obtained:

$$\underline{1}^T \mathbf{K}^{-1} \dot{\mathbf{E}} = -\left(\underline{1}^T \mathbf{L}\right)\left(\mathbf{Q}\chi^{-1}\right) = \mathbf{0}^T \left(\mathbf{Q}\chi^{-1}\right) = \mathbf{0} \tag{7.7}$$

And then (7.7) can be simplified as:

$$\sum_{i=1}^{n} \frac{\dot{E}_i}{k_i} = 0 \tag{7.8}$$

Moreover, the output voltage of each DG unit can be written as follows:

$$E_i = \int \left(\frac{dE_i}{d\tau}\right) d\tau = \int k_i \left(\frac{d\left(E_i / k_i\right)}{d\tau}\right) d\tau \tag{7.9}$$

Note that the average voltage of the microgrid distribution line is described as:

$$\bar{E}(t) = \frac{1}{n}\sum_{i=1}^{n} E_i \tag{7.10}$$

Therefore, combining with (7.8, 7.9, and 7.10), the averaged voltage magnitude can be written as:

$$\bar{E}(t) = \frac{1}{n}\sum_{i=1}^{n} E_i = \frac{1}{n}\sum_{i=1}^{n}\left[\int k_i \left(\frac{d\left(E_i / k_i\right)}{d\tau}\right) d\tau\right] = \frac{1}{n}\int k_i \left(\sum_{i=1}^{n}\frac{E_i}{k_i}\right)' d\tau = \frac{1}{n}\sum_{i=1}^{n} k_i E_i^d \tag{7.11}$$

where E_i^d is the initial voltage amplitude that can be regulated flexibly for practical application, that is, $Ed\ i$ is set to rated voltage amplitude, E, and k_i is equal to one, and the averaged voltage can be regulated as:

$$\bar{E}(t) = \frac{1}{n}\sum_{i=1}^{n} E_i = E^* \tag{7.12}$$

Hence, it can be concluded that the average voltage can be equal to the rated value as shown in (7.12).

3. *Active Power Sharing and Frequency Recovering*

As shown in Fig. 7.1, the active power of each DG is regulated by improved frequency droop controller without neighborhood active and frequency information. Equation (7.2) can be rewritten as (7.13) by using the equivalence between improved droop control and secondary control [12]:

$$\begin{cases} \omega_i = \omega_i^* - m_i P + \Delta\omega \\ \Delta\omega = k_{p,\omega i}\left(\omega^* - \omega_i\right) + \dfrac{k_{i,\omega i}}{s}\left(\omega^* - \omega_i\right) \end{cases} \tag{7.13}$$

where $k_{p,\omega i}$ and $k_{i,\omega i}$ are positive gains of DG_i, $\Delta\omega$ is a control variable for frequency compensation and P is the measured average active power through a low-pass filter.

In addition, the matrix format of (7.13) can be derived as:

$$\omega = \omega^* - \mathbf{m}\mathbf{P} + \mathbf{J}\left(\omega^* - \omega\right) \tag{7.14}$$

where $\mathbf{J} = \text{diag}\{k_{p\omega i} + k_{i\omega i}/s\}$, $i \in [1, N]$, and $k_{p\omega i}$ and $k_{i\omega i}$ are positive gains of DG_i. $\omega^* = \omega^* \mathbf{1}$, $\omega = [\omega\,1, \omega\,2,\ldots, \omega N]^T$, $\mathbf{m} = \text{diag}\{m_i\}$ and $\mathbf{P} = [P\,1, P\,2\ldots PN]^T$.

In the steady-state conditions, (7.14) can be rewritten as:

$$\omega^s = \omega^{s,*} - n\mathbf{P}^s + \mathbf{J}_p\left(\omega^{s,*} - \omega^s\right) + \mathbf{J}_I\left(\omega^{s,*} - \omega^s\right)\left(t - t_0\right) + \mathbf{K}_\omega\left(t_0\right) \tag{7.15}$$

where ω^s, $\omega^{s,*}$ and $\mathbf{P}^s$ represent the steady-state value of the ω, ω^* and $\mathbf{P}$, respectively. $\mathbf{J}_p$ and $\mathbf{J}_I$ are proportional and integral coefficients of $\mathbf{J}$, respectively.

Considering the time-dependent part of (7.15) is equal to zero in the steady state, the $\omega^s = \omega^{s,*}$ can be obtained. Note that the frequency is a global variable and active power sharing can be ensured by droop-based mechanical methods [10–14]. Therefore, the active power sharing can be realized while maintaining the frequency to the rated value for any DG unit.

7.3 Local Exponential Stability Analysis in Consensus Controller

In this section, the necessary and sufficient conditions for local exponential stability of equilibria of the system will be established for the islanded microgrid. Considering only the reactive power and voltage are regulated by consensus controller, to simplify complicated degree of stability analysis of the system and without loss of generality, the state equation of power stage with information of reactive power and voltage is established to analyze local exponential stability of the proposed consensus control scheme.

Note that the motion of an arbitrary voltage E_i can be expressed by all other voltages in (7.8, 7.9, 7.10, and 7.11), the local exponential stability analysis of the MG with dimension $2n$ can be simplified to study a reduced system with dimension $2n - 1$.

Considering the reactive power is defined as [39]:

$$Q_i = \underbrace{\left(\hat{B}_{ii} + \sum_{j \in N_i} B_{ij}\right)}_{B_{ii}} E_i^2 - \sum_{j \in N_i} B_{ij} E_i E_j \cos\theta_{ij}, \; Q_i = \frac{\omega_c}{s + \omega_c} q_i, \tag{7.16}$$

where q_i is instantaneous active power, N_i is the set of nodes connected to DG_i, $\hat{B}_{ii}$ is the negative of the shunt susceptance at node i, and B_{ij} is the negative of the suscep-tance between node i and node j.

By linearizing the voltage and frequency equations at the equilibrium point (E^s $i,Q^s i$) described in (7.5) and (7.16), and combining the partial derivative of the reac-tive power flow with respect to the rest of voltages, the following state equation can be derived:

$$
\begin{bmatrix} \dot{\tilde{E}}^s_{(n-1)\times1} & \dot{\tilde{Q}}^s_{n\times1} \end{bmatrix}^T = A_{sys} \begin{bmatrix} \tilde{E}^s_{(n-1)\times1} & \tilde{Q}^s_{n\times1} \end{bmatrix}^T \tag{7.17}
$$

$$
A_{sys} = \begin{bmatrix} 0 & -\begin{bmatrix} I_{n-1} & 0_{n-1} \end{bmatrix} KL\chi^{-1} \\ \omega_c N \underbrace{\begin{bmatrix} I_{n-1} \\ -\left(\dfrac{k_i}{k_1} \cdots \dfrac{k_i}{k_{i-1}}, \dfrac{k_i}{k_{i+1}} \cdots \dfrac{k_i}{k_n} \right)^T \end{bmatrix}}_{M} & -\omega_c I_n \end{bmatrix}
$$

where $\tilde{E}^s$ and $\tilde{Q}^s$ are represented by the equilibrium point E^s and Q^s with the small signal perturbation, respectively. And the elements of the matrix N are described as:

$$
n_{ii} = 2\left| \hat{B}_{ii} + \sum_{j\in N_i} B_{ij} \right| E_i^s - \sum_{N_i} \left| B_{ik} \right| E_k^s, \ n_{ik} = -\left| B_{ik} \right| E_i^s, i \neq k \tag{7.18}
$$

Let λ be an eigenvalue of A_{sys} with a corresponding right eigen vector $v = [v_1 v_2]^T$, the following equation can be derived:

$$
\begin{bmatrix} -\begin{bmatrix} I_{n-1} & 0_{n-1} \end{bmatrix} KL\chi^{-1}v_2 \\ \omega_c (NMv_1 - v_2) \end{bmatrix} = \lambda \begin{bmatrix} v_1 \\ v_2 \end{bmatrix} \tag{7.19}
$$

Equation (7.19) can be utilized to get the solution of the eigenvalue of A_{sys}. If the zero is an eigen value of A_{sys}, that is, $\lambda = 0$. Then, we get:

$$
-\begin{bmatrix} I_{n-1} & 0_{n-1} \end{bmatrix} KL\chi^{-1}v_2 = 0 \tag{7.20}
$$

It follows that (7.20) can only be satisfied if:

$$
KL\chi^{-1}v_2 = \begin{bmatrix} 0 & a \end{bmatrix}^T, \ a \in C \tag{7.21}
$$

Considering the *lemma* 1 that $\underline{1}^T_n L = \underline{0}^T_n$ together with $L = L^T$, implies that:

$$
\underline{1}^T_n K^{-1} KL\chi^{-1}v_2 = 0 \tag{7.22}
$$

Therefore:

$$\mathbf{1}_n^T \mathbf{K}^{-1} \mathbf{KL} \chi^{-1} v_2 = \mathbf{1}_n^T \mathbf{K}^{-1} \begin{bmatrix} 0 & a \end{bmatrix}^T = a / k_n = 0 \qquad (7.23)$$

Therefore, a must be zero in (7.23), which shows $\mathbf{KL} v_2 \chi^{-1} = \underline{0}_n$ and $v_2 = \beta \chi \mathbf{1}_n$, $\beta \in \mathbb{R}$ [31, 33]. Combining with (7.19), (7.24) can be obtained:

$$\mathbf{NM} v_1 = \beta \chi \underline{1}_n = \beta \kappa \mathbf{K}^{-1} \underline{1}_n \qquad (7.24)$$

It can be seen in (7.17) that $\mathbf{M}^T \mathbf{K}^{-1} \underline{1}_n = \underline{0}_{n-1}$, then $v*1 \mathbf{M}^T \mathbf{NM} v_1 = 0$ and $\mathbf{M} v_1 = \mathbf{0}_n$ can be derived from (7.24), where $v*1$ is the conjugate transpose of a vector v_1. Therefore, λ is equal to zero if and only if $v_1 = \mathbf{0}_{n-1}$ and $v_2 = \mathbf{0}_n$, which shows that zero is not an eigen value of $\mathbf{A}_{sys}$.

Since $\lambda \neq 0$, eigen values of (7.17) can be rewritten as:

$$\lambda^2 v_2 + \omega_c \lambda v_2 + \omega_c \mathbf{NM} \begin{bmatrix} \mathbf{I}_{n-1} & \mathbf{0}_{n-1} \end{bmatrix} \mathbf{KL} \chi^{-1} v_2 = \mathbf{0}_n \qquad (7.25)$$

Considering the real part of the numerical range of $\mathbf{N}$ is given by the range of its symmetric part of the numerical range of $1/2(\mathbf{N} + \mathbf{N}^T)$ [40], whose symmetric part has the following entries:

$$\begin{cases} \bar{n}_{ii} = n_{ii}, \bar{n}_{ik} = -\dfrac{1}{2} |B_{ik}| \left(V_i^s + V_k^s \right) \\ \bar{n}_{ii} = 2|B_{ii}| V_i^s - \sum_{k \in N_i} |B_{ik}| V_k^s > |B_{ii}| V_i^s > \sum_{k \in N_i, k \neq i} |\bar{n}_{ik}| \end{cases} \qquad (7.26)$$

Therefore, $\mathbf{N}$ is non-singular, the symmetric part is diagonally dominant with positive diagonal entries, and eigen values of $\mathbf{N}$ are positive real by Gershgorin's disc theorem [41].

Note that (7.27) can be obtained from (7.17):

$$\mathbf{M} \begin{bmatrix} \mathbf{I}_{n-1} & \mathbf{0}_{n-1} \end{bmatrix} \mathbf{KL} = \begin{bmatrix} \mathbf{I}_{n-1} & \mathbf{0}_{n-1} \\ -b^T & 0 \end{bmatrix} \mathbf{KL} = \mathbf{K} \begin{bmatrix} \mathbf{I}_{n-1} & \mathbf{0}_{n-1} \\ -\underline{1}_n^T & 0 \end{bmatrix} L = \mathbf{KL} \qquad (7.27)$$

And that:

$$\mathbf{M}^T \mathbf{K}^{-1} 1_{-n} = 0_{-n-1} \qquad (7.28)$$

Therefore, (7.29) can be derived:

$$\omega_c \mathbf{N} \left\{ \mathbf{M} \begin{bmatrix} \mathbf{I}_{n-1} \mathbf{0}_{n-1} \end{bmatrix} \mathbf{KL} \right\} \chi^{-1} v_2 = \omega_c \mathbf{N} \left\{ \mathbf{KL} \right\} \chi^{-1} v_2 = \omega_c \mathbf{N} \left\{ \chi^{-1} \mathcal{L} \right\} \chi^{-1} v_2 = \omega_c \gamma \mathbf{N} \chi^{-1} \mathcal{L} \chi^{-1} v_2 \quad (7.29)$$

where $\mathbf{K} = \gamma \chi^{-1}$ and γ is a positive real parameter.

Hence, (7.25) is equivalent to:

$$\lambda^2 v_2 + \omega_c \lambda v_2 + \omega_c \gamma \mathbf{N}\boldsymbol{\chi}^{-1}\mathbf{L}\boldsymbol{\chi}^{-1} v_2 = \mathbf{0}_n \tag{7.30}$$

In addition, the following equations can be derived by (7.17) and (7.29):

$$\mathbf{N}\boldsymbol{\chi}^{-1}\mathbf{L}\boldsymbol{\chi}^{-1} v = \mathbf{0}_n \Leftrightarrow \mathbf{L}\boldsymbol{\chi}^{-1} v = \mathbf{0}_n \Leftrightarrow v = \beta \boldsymbol{\chi} \mathbf{1}_n, \beta \in R, \beta \neq 0 \tag{7.31}$$

Therefore, $\mathbf{N}\boldsymbol{\chi}^{-1}\mathbf{L}\boldsymbol{\chi}^{-1}$ has a zero eigen value and a corresponding right eigen vector $\beta \boldsymbol{\chi} \mathbf{1}_n, \beta \neq 0$.

Note that the spectrum of $\mathbf{N}\boldsymbol{\chi}^{-1}\mathbf{L}\boldsymbol{\chi}^{-1}$ belongs to the numerical range of $\mathbf{N}$ multiply by the numerical range of $\boldsymbol{\chi}^{-1}\mathbf{L}\boldsymbol{\chi}^{-1}$ [36]. Considering $\boldsymbol{\chi}^{-1}\mathbf{L}\boldsymbol{\chi}^{-1}$ is positive semi-definite and then the numerical range of $\boldsymbol{\chi}^{-1}\mathbf{L}\boldsymbol{\chi}^{-1}$ belongs to a positive real number, and all non-zero eigen values of $\mathbf{N}\boldsymbol{\chi}^{-1}\mathbf{L}\boldsymbol{\chi}^{-1}$ have a positive real part.

Therefore, when $\mathbf{N}\boldsymbol{\chi}^{-1}\mathbf{L}\boldsymbol{\chi}^{-1} v_2 = \mathbf{0}_n$, the first non-zero eigen value $\lambda = -\omega_c$ of the matrix $\mathbf{A}$ can be derived.

In order to denote the remaining m, $m \in [0, 2n - 2]$, eigen values, considering complex eigen values of $\mathbf{N}\boldsymbol{\chi}^{-1}\mathbf{L}\boldsymbol{\chi}^{-1}$ by $\varphi_i = a_i + jb_i \in C$ with right eigen vector $\rho_i \in C$, that is, $\mathbf{N}\boldsymbol{\chi}^{-1}\mathbf{L}\boldsymbol{\chi}^{-1}\rho_i = \varphi_i \rho_i$, and $\rho_i^* \rho_i = 1$.

Therefore, the remaining m eigen values of $\mathbf{A}$ are the solutions $\lambda_{i1,2}$ of:

$$\lambda^2_{i1,2} + \omega_c \lambda_{i1,2} + \omega_c \gamma \varphi_i = 0 \tag{7.32}$$

If $b_i = 0$, all solutions of (7.32) have negative real parts by the Hurwitz condition [42]. If $b_i \neq 0$, the remaining m eigen values the matrix $\mathbf{A}$ of can be derived as:

$$\lambda_{i1,2} = \omega_c \left(-1 \pm \sqrt{1 - 4a_i\gamma / \omega_c + j(-4\gamma b_i / \omega_c)}\right) / 2 \tag{7.33}$$

According to $\sqrt{a + bi} = p + qi$, (7.33) is equivalent to:

$$\lambda_{i1,2} = -\frac{\omega_c}{2} + \frac{\omega_c}{2} \left(\frac{\sqrt{\sqrt{\left(1 - \frac{4a_i\gamma}{\omega_c}\right)^2 + \left(\frac{4\gamma b_i}{\omega_c}\right)^2} + \left(1 - \frac{4a_i\gamma}{\omega_c}\right)}\Big/ 2}{\pm i \sqrt{\sqrt{\left(1 - \frac{4a_i\gamma}{\omega_c}\right)^2 + \left(\frac{4\gamma b_i}{\omega_c}\right)^2} - \left(1 - \frac{4a_i\gamma}{\omega_c}\right)}\Big/ 2} \right) \tag{7.34}$$

Therefore, if and only if the following equation is satisfied:

$$\sqrt{\left[\sqrt{(1 - 4a_i\gamma / \omega_c)^2 + (4\gamma b_i / \omega_c)^2} + (1 - 4a_i\gamma / \omega_c)\right] / 2} < 1 \tag{7.35}$$

and both solutions $\lambda_{i1,2}$ in (7.32) have negative real parts. Hence, $\mathbf{A}$ is Hurwitz if and only if (7.35) holds for all φ_i, and then equilibrium point (*Es i*, *Q*ᵢ) of the system will be locally exponentially stable if and only if $\mathbf{A}$ is Hurwitz [42].

7.4   Results and Performance Evaluations

7.4.1   Comparison with the Conventional Droop and Secondary Control Strategies

To verify the effectiveness of the proposed consensus control scheme, an islanded MG, which consists of four parallel DG units, is implemented in Matlab/Simulink, with measurements recorded through a dSPACE 1006-based real-time digital simulator in this section, as shown in Fig. 7.2. Moreover, unequal feeder impedances and different power ratings of DG units are considered, where the nominal powers of DG_1 and DG_2 are rated at double capacity compared to DG_3 and DG_4. The complete parameters are given in Table 7.1.

As shown in Fig. 7.3a, initially, the loads are connected to the MG system and the steady-state active powers of each DG are 1000 W, 1000 W, 500 W and 500 W, respectively. At $t = 1.5$ s, the loads are disconnected from the MG system and the active powers of each DG drops to 750 W, 750 W, 380 W and 380 W, respectively. The largest fluctuation of active powers is about 30 W with the disturbance conditions of feeder and load; the system can be stable in a short regulation time (about 0.3 s). At $t = 3.5$ s, the loads are reconnected, the active powers can be restored to normal values within a small fluctuation and short regulation time, which indicates that the system can achieve active power sharing with load disturbance by the traditional droop control. However, due to the inherent defects of the traditional droop control, as shown in Fig. 7.3b–d, the reactive power, voltage and frequency deviations drop for 31.527 Var, 0.913 V and 0.158 Hz in the steady-state condition, respectively.

In order to restore the frequency and voltage amplitude to the rated values, a centralized secondary control strategy is applied in the system. As shown in Fig. 7.4, secondary controller is activated at $t = 1$ s and loads are disconnected at $t = 2.5$ s and reconnected at $t = 4$ s. As shown in Fig. 7.4a, no matter whether the secondary controller is activated or not, or in case of load disturbance, the active power of the system can be shared proportionally. In Fig. 7.4c, when the secondary controller is not activated, the voltage deviations of DG1 and DG2 are about 0.335 V and 0.315 V, and the voltage deviations of DG3 and DG4 are about 0.802 V and 0.904 V. After activation, the voltage amplitude is adjusted to the rated value of 325 V.

Compared with Fig. 7.3c, it is obvious that the deviation of the rated voltage is eliminated under load disturbance conditions. In Fig. 7.4d, the system frequency deviation is about 0.152 Hz at 0 ~ 1 s. After the secondary controller is activated, the system frequency is adjusted to the rated value of 50 Hz. Compared to Fig. 7.3d, the system frequency deviation is also eliminated and maintained at the rated value under load disturbance conditions. Thus, the centralized secondary control scheme shows better control performance than the conventional droop control. Although the secondary control overcomes the drawbacks of the traditional droop control, it still causes the deviation of reactive power due to the difference in line impedance [43,

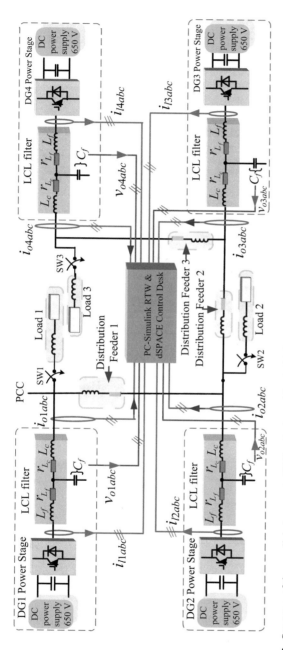

Fig. 7.2 Structure of the paralleled-connected DG units in an islanded AC MG

Table 7.1 Initial conditions and system parameters

System parameters	Values
LCL filter	$L_f = L_c = 1.8$ mH, $r_{Lf} = 0.15$ Ω, $r_{Lc} = 0.07$ Ω, and $C_f = 25$ μF
DC link voltage	650 V
DG feeder	$L_{line1} = 5.28$ mH $r_{line1} = 0.25$ Ω. $L_{line2} = 3.56$ mH $r_{line2} = 0.2$ Ω. $L_{line3} = 4.479$ mH $r_{line3} = 0.3$ Ω
Load	$L_{load1} = 0.182$H $r_{load1} = 200$ Ω. $L_{load2} = 0.462$H $r_{load2} = 120$ Ω. $L_{load3} = 0.251$H $r_{load3} = 85$ Ω
Voltage and current controllers	Values
$k_c, k_{pv}, k_{rv}, \omega_c, \omega_f$	$3, 0.175, 1, 5, 100\pi$
Control parameters	Values
Droop coefficients	$m_1 = m_2 = 0.0001$ rad/s/W, $m_3 = m_4 = 0.0002$ rad/s/W, $n_1 = n_2 = 0.0001$ V/Var, $n_3 = n_4 = 0.0002$ V/Var
Consensus controller	Values
ω_h, b, k_p, k_i	$4\pi, 1.2, 0.01, 0.24$
DG1- $k_{p,\omega i}, k_{i,\omega i}$	0.8, 10
DG2- $k_{p,\omega i}, k_{i,\omega i}$	0.76, 8.68
DG3- $k_{p,\omega i}, k_{i,\omega i}$	0.66, 9.24
DG4- $k_{p,\omega i}, k_{i,\omega i}$	0.68, 10.2

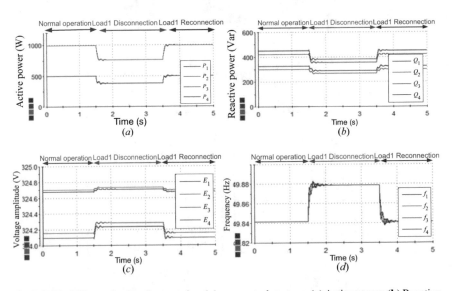

Fig. 7.3 The HIL test results of conventional droop control strategy. (**a**) Active power. (**b**) Reactive power. (**c**) Voltage amplitude. (**d**) Frequency

44]. This is also due to the intrinsic defects of the secondary control mechanisms. When the secondary control restores the voltage of the PCC point between the DG units to the rated value, the reactive power of each DG unit cannot be effectively shared, the difference is even larger. Similarly, if the reactive powers of each DG

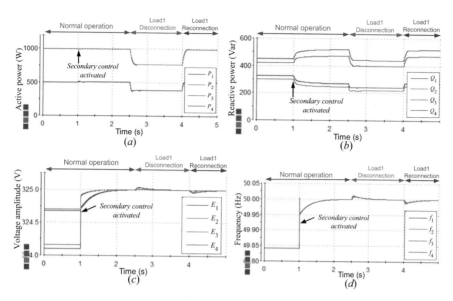

Fig. 7.4 The HIL test results of conventional secondary control strategy. (**a**) Active power. (**b**) Reactive power. (**c**) Voltage amplitude. (**d**) Frequency

unit are shared equally, the PCC voltage of each DG unit cannot be recovered to the rated value and is even more deviated than the case when only the droop control is applied [45, 46].

Therefore, compared to Fig. 7.3b, the reactive power deviation of the system under the conventional droop control is 31.527 Var, while the reactive power deviation of the system under the secondary control is approximately 47.761 Var, as shown in Fig. 7.4b. The reactive power deviation is obviously larger than that of the conventional droop control strategy. Due to the difference of line impedances and load impedances among DG units, the reactive power deviation of the system is unavoidable, which indicates that the poor reactive power regulation still exists in the system.

The dynamic response of the proposed consensus-based enhanced droop controller, with disturbance of loads, is shown in Fig. 7.5. In this scenario, the active and reactive powers can be shared proportionally while regulating average voltage to the rated value, as shown in Fig. 7.5a, b. Note that there is only a small dynamic fluctuation in the reactive power (less than 15.517 Var) with the disturbance conditions of feeder and load, as shown in Fig. 7.5b.

It is suggested that individual bus voltages should slightly deviate from the rated value (less than 5%) [30], to achieve accurate reactive power sharing of each DG unit. As shown in Fig. 7.5c, the largest voltage difference is less than 0.09 V (about 0.15% of the rated voltage amplitude), when sharing the active and reactive powers of the islanded MG system. Moreover, Fig. 7.5d shows that the frequency can be restored to the rated value without neighborhood voltage and frequency information. Additionally, the smaller fluctuation (about 0.028 Hz) and less regulation time

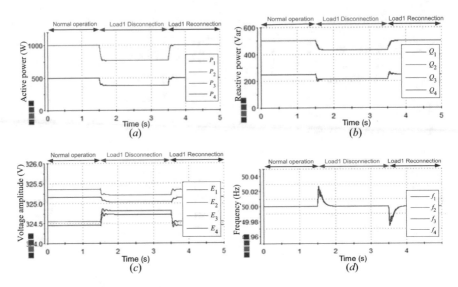

Fig. 7.5 The HIL test results of the consensus-based enhanced droop control. (**a**) Active power. (**b**) Reactive power. (**c**) Voltage amplitude. (**d**) Frequency

(of about 0.51 s) can be seen in Fig. 7.5, compared to the traditional control strategies in Figs. 7.3 and 7.4.

In addition, the influence of the communication delay on the proposed control approach was also investigated and a continuous 400 ms delay was added to the DG low-bandwidth sparse communication network of the system. The improved frequency controller is immune to time delay as no neighborhood active power and frequency information is required in the proposed scheme. Therefore, the accurate reactive power sharing and frequency recovery can be realized in case of the communication delay or even communication system crashes. Therefore, it can be seen from Figs. 7.6a and d that the active power sharing and frequency recovery of the system are the same as those without a communication delay (as shown in Fig. 7.5a and d). Although a small fluctuation of output voltage can be seen in Fig. 7.6c, it can be adjusted in a short time (less than 1 s). As shown in Fig. 7.6b, even in case of communication delay, accurate reactive power sharing is not affected, and the reactive power sharing result is basically the same as the case without delay, which shows the high reliability of the MG system.

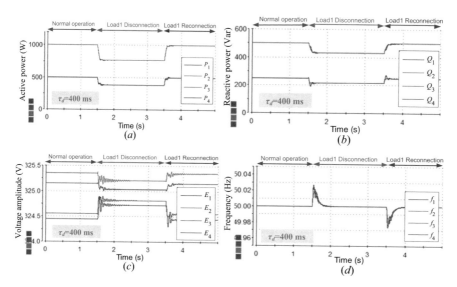

Fig. 7.6 The HIL test results of the consensus-based enhanced droop control considering time delay. (**a**) Active power. (**b**) Reactive power. (**c**) Voltage amplitude. (**d**) Frequency

7.4.2   Performance Evaluations in Case of Communication Failure

In order to validate the effectiveness of the proposed consensus control scheme in case of communication failure, the dynamic response process of the conventional secondary control and the conventional consensus control scheme are also presented, for the sake of comparison. Note that the unequal feeder impedances and different power ratings of DG units are considered, where the nominal powers of DG1 and DG2 are rated at the double capacity compared to DG3 and DG4, and the values of system parameters are consistent with the case in Table I.

Figure 7.7 shows the dynamic response of the conventional secondary control. Only droop controller is effective at the beginning and the secondary controller is activated at $t = 1$ s. Although voltage amplitude and frequency of the system can be regulated to the rated value, the poor reactive regulation still exists in the system. The communication lines are disconnected at $t = 5$ s, voltage amplitude and frequency return to the initial value, especially the voltage deviations and frequency deviations of DG1-DG4 drop for 0.342 V, 0.324 V, 0.712 V, 0.824 V and 0.156 Hz, respectively. Note that the accurate active power sharing of the conventional secondary scheme can also be realized under communication failures.

Figure 7.8 shows the dynamic response process of the conventional consensus control Scheme [30]. In this scenario, before the communication lines are disconnected, the active and reactive powers sharing can be realized effectively, and frequency can also be regulated to the rated value. At $t = 5$ s, the communication line is disconnected to emulate the communication failure. Figures 7.8a, b show that the

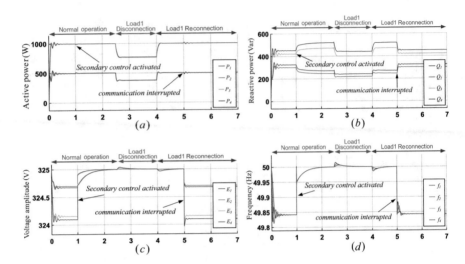

Fig. 7.7 The simulation results of conventional secondary control strategy. (**a**) Active power. (**b**) Reactive power. (**c**) Voltage amplitude. (**d**) Frequency

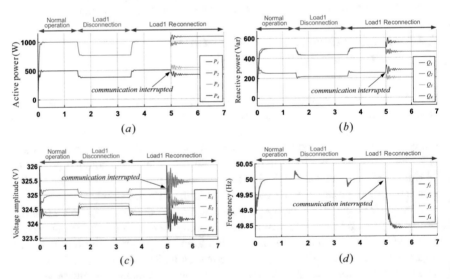

Fig. 7.8 The simulation results of the conventional consensus control. (**a**) Active power. (**b**) Reactive power. (**c**) Voltage amplitude. (**d**) Frequency

active and reactive powers cannot be shared proportionally, that is, the deviations of active and reactive powers reach to 112.241 W and 90.356 Var. Moreover, the large fluctuations of output voltage can be seen in Fig. 7.8c and the voltage deviations become larger in the steady-state condition, with the frequency dropping to 49.84 Hz. The simulation results show that the MG system, which the conventional consensus control scheme cannot realize, is accurate, active and reactive power

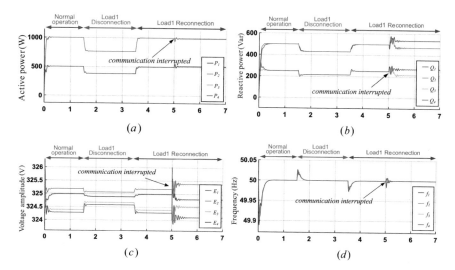

Fig. 7.9 The simulation results of the consensus-based enhanced droop control. (**a**) Active power. (**b**) Reactive power. (**c**) Voltage amplitude. (**d**) Frequency

sharing, and has voltage regulation and frequency recovery in case of communication failures.

Figure 7.9 shows the performance of the proposed control scheme in case of communication failures and load disturbances. Initially, the active and reactive powers sharing, frequency recovery can be realized effectively. Note that the voltages slightly deviate from the rated value (less than 5%) in order to achieve accurate reactive power sharing. The communication lines are interrupted at $t = 1$ s. Figure 7.9a and d demonstrate the proportional active power sharing and frequency recovery. Referring to Fig. 7.7a and d, it can be observed that the conventional secondary control scheme can achieve active power sharing, but the frequency recovery cannot be realized. Hence, the proposed control scheme shows a better performance than conventional secondary controller in case of communication failures. Moreover, Fig. 7.8a and d shows that the conventional consensus controller cannot achieve active power sharing and frequency recovery under communication failures. Therefore, by using the consensus-based enhanced droop control scheme, the high reliability of the MG system can be ensured in case of load disturbances, communication delay, and even in case of communication failures.

7.5   Conclusion

This chapter proposes a consensus-based enhanced droop control scheme in islanded microgrid system. Compared to conventional droop and secondary control methods, the proposed control scheme can share reactive power proportionally while regulating frequency and average voltage to the rated values, maintaining output voltages of each DG within an acceptable range and improving dynamic performance of the system. Different from the existing consensus control methods, only neighborhood reactive power information is required, and the transfer data and data latency are significantly reduced, and the high reliability of the system can be achieved. Moreover, the active power sharing, frequency recovering and stability of the system can be ensured even in case of a total communication link failure.

Moreover, the steady-state performance analysis of the proposed consensus control is presented in this chapter, including reactive power sharing, average voltage regulation, active power sharing and frequency recovering. In addition, the necessary and sufficient conditions for local exponential stability of the equilibrium part of the system are established to validate the proposed consensus-based enhanced droop control method. Finally, the various case studies are presented to show that the accurate active and reactive power sharing, and voltage and frequency regulation, can be realized in the proposed consensus controller in case of load disturbances, communication delay or in case of communication failures.

The communication burden of systems is significantly reduced and the robustness under time delay and the reliability of systems are greatly enhanced, which further validates the effectiveness of the proposed approach for islanded MG systems. Notably, the proposed consensus control is mainly applicable to micro-grid systems with inductive impedance as main line impedance when the output inductor is used and/or the possible presence of an output transformer. Under these circumstances, the inductive parts dominate the resistive parts in the admittances for some particular microgrids, especially on the MV level [47, 48]. In case of the pure resistive or capacitive line impedance conditions of the MG system, some further in-depth research needs to be performed, which would be further studied in the near future.

References

1. Sun, Q., Guerrero, J. M., Jing, T., Vasquez, J. C., & Yang, R. (2017). An islanding detection method by using frequency positive feedback based on FLL for single-phase microgrids. *IEEE Transactions on Smart Grid, 5*(3), 982–994.
2. Shuai, Z., Hu, Y., Peng, Y., Tu, C., & Shen, Z. J. (2017). Dynamic stability analysis of synchronverter-dominated microgrid based on bifurcation theory. *IEEE Transactions on Industrial Electronics, 64*(9), 7467–7477.
3. Bani-Ahmed, A., Rashidi, M., Nasiri, A., & Hosseini, H. Reliability analysis of a decentralized microgrid control architecture. *IEEE Transactions on Smart Grid*, to be published. https://doi.org/10.1109/TSG.2018.2843527
4. Shuai, Z., Huang, W., Shen, X., Li, Y., Zhang, X., & Shen, Z. J. A maximum power loading factor (MPLF) control strategy for distributed secondary frequency regulation of islanded microgrid. *IEEE Transactions on Power Electronics*, to be published. https://doi.org/10.1109/TPEL.2018.2837125
5. Lasseter, R. H. (2002). Microgrids. In *Proc. IEEE Power Engineering Society Winter Meeting* (pp. 305–308).

6. Hatziargyriou, N., Asano, H., Iravani, R., & Marnay, C. (2007). Microgrids. *IEEE Power and Energy Magazine, 5*(4), 78–94.
7. Guerrero, J. M., Vasquez, J. C., Matas, J., Vicuna, L. G. d., & Castilla, M. (2011). Hierarchical control of droop-controlled AC and DC microgrids—A general approach toward standardization. *IEEE Transactions on Industrial Electronics, 28*(1), 158–172.
8. Lopes, J. A. P., Moreira, C. L., & Madureira, A. G. (2006). Defining control strategies for microgrids islanded operation. *IEEE Transactions on Power Systems, 21*(2), 916–924.
9. Olivares, D. E., et al. (2014). Trends in microgrid control. *IEEE Transactions on Smart Grid, 5*(4), 1905–1911.
10. Sun, Y., Hou, X., Yang, J., Han, H., Su, M., & Guerrero, J. M. (2017). New perspectives on droop control in AC microgrid. *IEEE Transactions on Industrial Electronics, 64*(7), 5741–5745.
11. Guerrero, J. M., Vasquez, J. C., Matas, J., Vicuna, L. G. D., & Castilla, M. (2011). Hierarchical control of droop-controlled AC and DC microgrids-general approach toward standardization. *IEEE Transactions on Industrial Electronics, 58*(1), 158–172.
12. Han, Y., Li, H., Xu, L., Zhao, X., & Guerrero, J. M. (2018). Analysis of washout filter-based power sharing strategy—An equivalent secondary controller for islanded microgrid without LBC lines. *IEEE Transactions on Smart Grid, 9*(5), 4061–4076.
13. Gu, Y. J., Li, W. H., & He, X. N. (2015). Frequency-coordinating virtual impedance for autonomous power management of DC microgrid. *IEEE Transactions on Power Electronics, 30*(4), 2328–2337.
14. Sreekumar, P., & Khadkikar, V. (2016). A new virtual harmonic impedance scheme for harmonic power sharing in an islanded microgrid. *IEEE Transactions on Power Delivery, 31*(3), 936–945.
15. Guerrero, J. M., Chandorkar, M., Lee, T., & Loh, P. C. (2013). Advanced control architectures for intelligent microgrids-part I: Decentralized and hierarchical control. *IEEE Transactions on Industrial Electronics, 60*(4), 1254–1262.
16. Mahmood, H., Michaelson, D., & Jiang, J. (2015). Accurate reactive power sharing in an islanded microgrid using adaptive virtual impedances. *IEEE Transactions on Power Electronics, 30*(3), 1605–1617.
17. Yuen, C., Oudalov, A., & Timbus, A. (2011). The provision of frequency control reserves from multiple microgrids. *IEEE Transactions on Industrial Electronics, 58*(1), 173–183.
18. Lee, C. T., Chu, C. C., & Cheng, P. T. (2013). A new droop control method for the autonomous operation of distributed energy resource interface converters. *IEEE Transactions on Power Electronics, 28*(4), 1980–1993.
19. Mehrizi-Sani, A., & Iravani, R. (2010). Potential-function based control of a microgrid in islanded and grid-connected models. *IEEE Transactions on Power Apparatus and Systems, 25*(4), 1883–1891.
20. Lopes, J. A. P., Moreira, C. L., & Madureira, A. G. (2006). Defining control strategies for microgrids islanded operation. *IEEE Transactions on Power Apparatus and Systems, 21*(2), 916–924.
21. Bidram, A., Davoudi, A., & Lewis, F. L. (2014). A multi-objective distributed control framework for islanded AC microgrids. *IEEE Transactions on Industrial Informatics, 10*(3), 1785–1798.
22. Han, R., Meng, L., Trecate, G. F., Coelho, E. A. A., Vasquez, J. C., & Guerrero, J. M. Containment and consensus-based distributed coordination control to achieve bounded voltage and precise reactive power sharing in islanded AC microgrids. *IEEE Transactions on Industry Applications*, to be published. https://doi.org/10.1109/TIA.2017.2733457
23. Dou, C., Zhang, Z., Yue, D., & Zheng, Y. (2017). MAS-based hierarchical distributed coordinate control strategy of virtual power source voltage in low-voltage microgrid. *IEEE Access, 5*, 11381–11390.

24. Lü, P., Zhao, J., Yao, J., & Yang, S. (2017). A decentralized approach for frequency control and economic dispatch in smart grids. *IEEE Journal on Emerging and Selected Topics in Circuits and Systems, 7*(3), 447–458.
25. Guan, Y., Meng, L., Li, C., Vasquez, J. C., & Guerrero, J. M. A dynamic consensus algorithm to adjust virtual impedance loops for discharge rate balancing of AC microgrid energy storage units. *IEEE Transactions on Smart Grid*, to be published. https://doi.org/10.1109/TSG.2017.2672882
26. Xu, Y., & Li, Z. (2015). Distributed optimal resource management based on the consensus algorithm in a microgrid. *IEEE Transactions on Industrial Electronics, 62*(4), 2584–2592.
27. Hao, R., Jiang, Z., Ai, Q., Yu, Z., & Zhu, Y. (2018). Hierarchical optimisation strategy in microgrid based on the consensus of multiagent system. *IET Generation Transmission and Distribution, 12*(10), 2444–2451.
28. Zheng, Y., Song, Y., Hill, D. J., & Zhang, Y. (2018). Multiagent system based microgrid energy management via asynchronous consensus ADMM. *IEEE Transactions on Energy Conversion, 33*(2), 886–888.
29. Chen, G., & Zhao, Z. (2018). Delay effects on consensus-based distributed economic dispatch algorithm in microgrid. *IEEE Transactions on Power Systems, 33*(1), 602–612.
30. Nasirian, V., Shafiee, Q., Guerrero, J., Lewis, F., & Davoudi, A. (2016). Droop-free distributed control for AC microgrids. *IEEE Transactions on Power Electronics, 31*(2), 1600–1161.
31. Zhang, H., Kim, S., Sun, Q., & Zhou, J. (2017). Distributed adaptive virtual impedance control for accurate reactive power sharing based on consensus control in microgrids. *IEEE Transactions on Smart Grid, 8*(4), 1749–1761.
32. Shafiee, Q., Nasirian, V., Vasquez, J. C., Guerrero, J. M., & Davoudi, A. A multi-functional fully distributed control framework for AC microgrids. *IEEE Transactions on Smart Grid*, to be published. https://doi.org/10.1109/TSG.2016.2628785
33. Lu, L. Y., & Chu, C. C. (2015). Consensus-based droop control synthesis for multiple DICs in isolated micro-grids. *IEEE Transactions on Power Apparatus and Systems, 30*(5), 2243–2256.
34. Zhou, J., Kim, S., Zhang, H., Sun, Q., & Han, R. Consensus-based distributed control for accurate reactive, harmonic and imbalance power sharing in Microgrids. *IEEE Transactions on Smart Grid*, to be published. https://doi.org/10.1109/TSG.2016.2613143
35. Efimov, D., Polyakov, A., Perruquetti, W., & Richard, J. P. (2016). Weighted homogeneity for time-delay systems: Finite-time and independent of delay stability. *IEEE Transactions on Automatic Control, 55*(12), 210–215.
36. Li, Z., & Chen, J. (2017). Robust consensus of linear feedback protocols over uncertain network graphs. *IEEE Transactions on Automatic Control, 62*(8), 4251–4258.
37. Schiffer, J., Seel, T., Raisch, J., & Sezi, T. (2016). Voltage stability and reactive power sharing in inverter-based microgrids with consensus-based distributed voltage control. *IEEE Transactions on Control Systems Technology, 24*(1), 96–109.
38. Huang, J. (2017). The consensus for discrete-time linear multi-agent systems under directed switching networks. *IEEE Transactions on Automatic Control, 62*(8), 4086–4092.
39. Persis, C. D., & Monshizadeh, N. Bregman storage functions for microgrid control. *IEEE Transactions on Automatic Control*, to be published. https://doi.org/10.1109/TSG.2017.2647958
40. Horn, R. A., & Johnson, C. R. (1991). *Topics in matrix analysis.* Cambridge University Press.
41. Horn, R. A., & Johnson, C. R. (2012). *Matrix analysis.* Cambridge University Press.
42. Khalil, H. K. (2002). *Nonlinear systems* (Vol. 3). Prentice-Hall.
43. Rocabert, J., Luna, A., Blaabjerg, F., & Rodríguez, P. (2012). Control of power converters in AC microgrids. *IEEE Transactions on Power Electronics, 27*(11), 4734–4749.
44. Bidram, A., & Davoudi, A. (2012). Hierarchical structure of microgrids control system. *IEEE Transactions on Smart Grid, 3*(4), 1963–1976.
45. Guerrero, J. M., Matas, J., de Luis Garcia, V., Castilla, M., & Miret, J. (2007). Decentralized control for parallel operation of distributed generation inverters using resistive output impedance. *IEEE Transactions on Industrial Electronics, 54*, 994–1004.

46. Jaehong, K., Guerrero, J. M., Rodriguez, P., Teodorescu, R., & Kwanghee, N. (2011). Mode adaptive droop control with virtual output impedances for an inverter-based flexible AC microgrid. *IEEE Transactions on Power Electronics, 26*, 689–701.
47. Simpson-Porco, J. W., Dörfler, F., & Bullo, F. (2013). Synchronization and power sharing for droop-controlled inverters in islanded microgrids. *Automatica, 49*(9), 2603–2611.
48. Schiffer, J., Ortega, R., Astolfi, A., Raisch, J., & Sezi, T. (2014). Conditions for stability of droop-controlled inverter-based microgrids. *Automatica, 50*(10), 2457–2469.

Chapter 8
Enhanced Hierarchical Control for Islanded Microgrid Using Advanced Damping Methods

In this chapter, the modeling, controller design, and stability analysis of the islanded microgrid (MG) using enhanced hierarchical control structure with multiple current loop damping schemes is proposed. The islanded MG consists of the parallel-connected voltage source inverters using LCL output filters, and the proposed control structure includes: the primary control with additional phase-shift loop, the secondary control for voltage amplitude and frequency restoration, the virtual impedance loops which contains virtual positive- and negative-sequence impedance loops at fundamental frequency, and virtual variable harmonic impedance loop at harmonic frequencies, and the inner voltage and current loop controllers. A small-signal model for the primary and secondary controls with additional phase-shift loop is presented, which shows an overdamped feature from eigenvalue analysis of the state matrix. The moving average filter-based sequence decomposition method is proposed to extract the fundamental positive and negative sequences, and harmonic components. The multiple inner current loop damping scheme is presented, including the virtual positive, virtual negative, and variable harmonic sequence impedance loops for reactive and harmonic power sharing purposes and proposed active damping (AD) scheme using capacitor current feedback loop of the LCL filter, which shows enhanced damping characteristics and improved inner-loop stability features. The simulation results obtained from EMTP under nonidentical line impedance scenario are presented, and the effect of the low-bandwidth communication (LBC) delay is also simulated and compared with ideal scenario. Finally, the experimental results are also provided to validate the feasibility of the proposed approach, which can be widely applied in practical applications.

8.1   Introduction

Microgrids (MGs), which contain a number of systematically organized DG units, have been emerging as a framework to overcome the problems caused by the high levels of penetration of DG and make large-scale application of DG possible. As an interface between the DG and the power grid or local loads, a voltage source inverter (VSI) is the most common topology which can operate either in grid-connected or islanded mode to provide a controlled and high-quality power exchange with the grid or local loads. In islanded mode, the local loads should be supplied by the DG units, which now act as the controlled voltage sources (CVS). And the MGs need some form of control in order to avoid circulating currents between the DG units and ensure stable and efficient operation. The important roles that can be achieved using these control structures are active and reactive power control capabilities among the DGs, energy management, frequency and voltage regulation, and economic optimization [1–3].

Many control strategies of parallel VSIs forming an MG have been investigated, where the decentralized and cooperative controllers such as the droop methods have been proposed and are considered as preferred option due to several attractive advantages such as flexibility, and no need of high bandwidth communications. The droop control is a kind of cooperative control that allows the parallel connection of VSIs sharing active and reactive powers. In order to analyze the performance of these methods, the state-space and small-signal models can be utilized to address the stability and dynamic performance of the droop-controlled MGs. The virtual impedance can be added in the control loop to enhance the reliability and performance of the droop-controlled VSIs, ensuring the inductive behavior of the output impedance of the DGs. The transients during fast load switching and voltage quality under nonlinear loads in islanded MG may also be influenced by the virtual impedance effects [4, 5].

It is well-known that the droop control methods involve an inherent trade-off between power sharing and voltage and frequency regulation. On the other hand, the secondary controllers can be used to compensate the deviations of voltage and frequency. The islanded MGs can restore the frequency and voltage amplitude in spite of deviations created by the total amount of active and reactive powers demanded by the loads. Moreover, the increasing proliferation of nonlinear loads could result in significant harmonic distortion in the distribution systems. And an MG should be able to operate under nonlinear load conditions without performance degradations. Based on the IEEE standard 519–1992 [6], the voltage total harmonic distortion (THD) for sensitive loads should be maintained below 5%. In industrial applications, LC filters are usually used as an interface between the inverter and the local loads to effectively mitigate the harmonic contents of the inverter output waveforms. The pure LC or LCL filters are highly susceptible to resonances with harmonic components generated by the inverter or the distorted nonlinear loads. In order to mitigate system resonances, a damping resistor can be placed in the LC or LCL filters, but it results in power losses. To avoid drawbacks of passive damping, various active damping (AD) methods based on the inner loop feedback variables can be applied. Among AD methods, the method involving feedback of the capacitor current of the LCL filter has attracted plenty of attention due to its effectiveness, simple implementation, and wide application [7].

On the other hand, the various hierarchical control strategies for MGs have been presented in [8–12], where the power oscillations, accuracy of the droop control, and the power sharing problems are seldom considered. In [8], the parallel-connected bidirectional converters for AC and DC hybrid MG application are analyzed in standalone operation mode and the conventional hierarchical control in stationary frame under resistive conditions is adopted. A general approach of hierarchical control for MG is presented in [9]; the tertiary control could provide high-level inertias to interconnect more MGs, acting as the primary control of the cluster, and the tertiary cluster control can fix the active and reactive powers to be provided by this cluster or act like a primary control to interconnect more MG clusters. In another hierarchical control structure, an autonomous active power control strategy is modeled for MGs with photovoltaic (PV) cell generation, energy storage system (ESS), and loads to achieve power management in a decentralized manner [10], and the state of charge (SoC) of the ESS can be kept within the safe limits by automatically adjusting the power generation from the PV systems and load consumption. In [11], the coordinated control of DG inverter and active power filter (APF) to compensate voltage harmonics in MG is addressed, the APF participates in harmonic compensation and consequently the compensation efforts of DG decrease to avoid excessive harmonics or loading of the DG inverter. In [12], the advanced decentralized and hierarchical control methods are reviewed, and the future trends in hierarchical control for MGs and the clusters of MGs are given.

In this chapter, the enhanced hierarchical control for islanded MG is presented, using multiple inner loop active damping schemes, which shows improved characteristics in terms of hierarchical control structure including the droop control with additional phase-shift loop and the centralized secondary control scheme for voltage amplitude and frequency restoration purposes of MG. A small signal model is developed for the power control loop, which takes the droop control with additional phase-shift loop and secondary control into consideration. The power control loop is designed to be overdamped to suppress power oscillations. The proposed multiple active damping and virtual impedance schemes are adopted for the inner loops, that is, the capacitor current feedback loop plus the output virtual impedance loops, which achieves the purposes of inverter side LC resonance active damping, reactive power, and harmonic power sharing.

The proposed approach employs the moving average filter-based sequence decomposition which is composed of the virtual positive- and negative-sequence impedance loops at fundamental frequency, and the virtual variable harmonic impedance loop at harmonic frequencies [13, 14]. The virtual positive- and negative-sequence impedance loops improve the performance of the active power-frequency (P-ω) and reactive power-voltage magnitude (Q-E) droop controllers and reduce the fundamental negative sequence circulating current. A proper sharing of harmonic power among all the DG inverters is achieved by using the virtual variable harmonic impedance loop at characteristic harmonic frequencies. The feasibility of the proposed approach is first evaluated using EMTP-based time domain simulation studies under various operating scenarios, and then validated by the experimental results obtained from two reduced-scale parallel-connected PWM inverters under linear and nonlinear load conditions.

The remainder of this chapter is structured as follows. The enhanced hierarchical control structure and strategy of the islanded MG system are analyzed in detail in Sects. 8.2 and 8.3, including the droop control, secondary control, small-signal analysis of the power controller, virtual impedance loop, inner voltage and current control loops. The simulation results obtained from EMTP under nonidentical line impedance scenario are presented in Sect. 8.4, and the effect of the low-bandwidth communication (LBC) delay is also considered. Section 8.5 provides comprehensive experimental results. Finally, Sect. 8.6 concludes this chapter.

8.2   Enhanced Hierarchical Control Strategy

A typical structure of MG with n DGs and loads is given in Fig. 8.1. Although the proposed control strategies can operate in either the grid-connected mode or islanded mode, only the islanded operation mode will be considered in this chapter. The three-phase VSIs with LCL filters are usually used as the DG interfaces to connect with the local AC bus, and the power stage of two DGs and the proposed control strategy for their interface inverters connected in an islanded mode are shown in Fig. 8.2. Each DG unit with its LCL filter can be considered as a subsystem of the MG.

As shown in Fig. 8.3, the control strategy of an individual DG unit is implemented in the stationary reference frame. The dynamics of the DG units are influenced by the output LCL filter, the droop controller, the secondary controller with frequency and voltage amplitude restoration, the power calculation, the virtual impedance loops which contain the virtual positive- and negative-sequence impedance loops at fundamental frequency, and the virtual variable harmonic impedance loop at harmonic frequencies, and the PR-based inner voltage and current loops [15, 16].

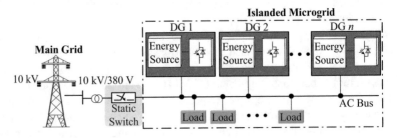

Fig. 8.1 Typical structure of MG with DGs and multiple loads

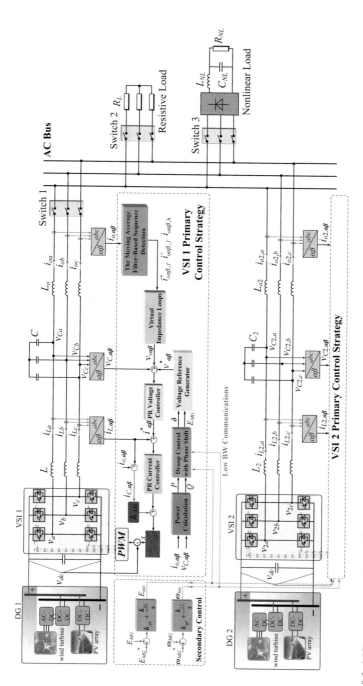

Fig. 8.2 MG power stage and control system

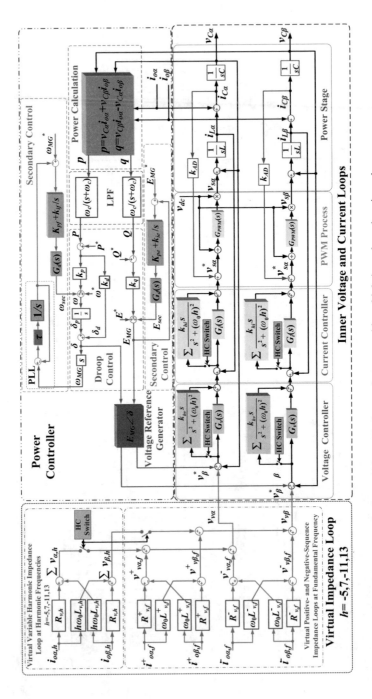

Fig. 8.3 Block diagram of the proposed enhanced hierarchical control strategy with multiple inner-loop damping schemes

8.2.1   Droop Control for Microgrid

The droop control is utilized to avoid communication wires while obtaining good power sharing, which is responsible for adjusting the frequency and amplitude of the voltage reference according to the active and reactive powers (P and Q), ensuring P and Q flow control. The active power frequency (P-ω) and reactive power voltage magnitude (Q-E) droop control schemes are defined as

$$\omega = \omega^* - k_p\left(P - P^*\right), \quad E = E^* - k_q\left(Q - Q^*\right) \tag{8.1}$$

where ω and E represent the frequency and amplitude of the output voltage references, ω^* and E^* are the nominal frequency and amplitude, P^* and Q^* are the active and reactive power references normally set to zero in islanded MG, and k_p and k_q are the droop coefficients.

Referring to Ref. [4], the instantaneous active power (p) and reactive power (q) are calculated from the $\alpha\beta$-axis output voltage ($v_{C\alpha\beta}$) and current ($i_{o\alpha\beta}$) as:

$$p = v_{C\alpha}i_{o\alpha} + v_{C\beta}i_{o\beta}, \quad q = v_{C\beta}i_{o\alpha} - v_{C\alpha}i_{o\beta} \tag{8.2}$$

The instantaneous powers are then passed through low-pass filters with the cut-off frequency ω_c to obtain the filtered output real and reactive powers (P and Q) as follows:

$$P = \frac{\omega_c}{s + \omega_c}p, \quad Q = \frac{\omega_c}{s + \omega_c}q \tag{8.3}$$

The bandwidth of the low pass filter is much smaller than that of the inner controllers of DG unit and the performance of the system is strongly influenced by this fact.

8.2.2   Secondary Control for Microgrid

The inherent trade-off between power sharing and voltage and frequency regulation is one drawback of the droop method. The conventional droop control is local and does not have communications with other DG units. In order to mitigate these disadvantages, a restoration control can be added to remove any steady-state error introduced by the conventional droop and achieve global controllability of the MG that ensures nominal values of voltage and frequency in the MG.

As shown in Fig. 8.2, the primary and secondary controls are implemented in each DG unit. The secondary control is realized by low bandwidth communication among the DG units. By using this approach, the frequency and voltage amplitude restoration compensators can be derived as:

$$\begin{cases} \omega_{sec} = k_{pf}\left(\omega_{MG}^* - \omega_{MG}\right) + k_{if}\int\left(\omega_{MG}^* - \omega_{MG}\right)dt \\ E_{sec} = k_{pe}\left(E_{MG}^* - E_{MG}\right) + k_{ie}\int\left(E_{MG}^* - E_{MG}\right)dt \end{cases} \tag{8.4}$$

where k_{pf}, k_{if}, k_{pe}, and k_{ie} are the control parameters of the proportional integral (PI) compensator of the frequency and voltage restoration control, respectively. The angular frequency levels in the MG (ω_{MG}) are measured and compared to the reference ($\omega^* MG$) and the errors processed by the PI compensator are sent to all the DGs in order to restore the frequency of MG. The control signal (E_{sec}) is sent to the primary control level of each DG in order to remove the steady-state errors of the droop control.

1. *Frequency control*: Taking the idea from large electrical power systems, in order to compensate the frequency deviation produced by the local P-ω droop controllers, secondary frequency controllers have been proposed in Ref. [4]. A model of the frequency secondary control is shown in Fig. 8.4, which is also depicted in Fig. 8.3 in detail.

The control block diagram in Fig. 8.4 includes the droop control and secondary control. For droop control model, a low-pass filter ($G_{LPF}[s]$) with cutting frequency of 5 Hz has been considered for power calculation. The secondary control has been modeled by means of a simplified PLL first-order transfer function ($G_{PLL}[s]$) with the gain (τ) used to extract the frequency of the MG, the secondary PI controller ($G_{fsec}[s]$) is used to restore the frequency deviations, and a proportional gain (k_d) of the additional phase-shift loop is super-imposed to the active power control loop to suppress power oscillation. The additional phase-shift loop performs a phase displacement δ_d which is added to the phase determined by the P-ω droop δ_p, resulting in the angle δ of the inverter voltage E_{MG}. This strategy increases system damping, and the phase of inverter δ which is used by the reference generator block, calculated by

$$\delta = \delta_p + \delta_d \tag{8.5}$$

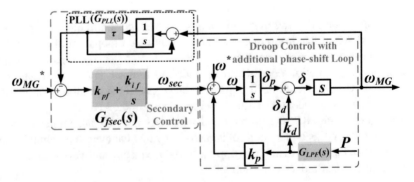

Fig. 8.4 Block diagram of frequency control for a DG unit

From the block diagram of Fig. 8.4, ω_{MG} is derived as

$$\omega_{MG} = \frac{G_{f\,sec}(s)}{1+G_{f\,sec}(s)G_{PLL}(s)}\omega_{MG}^{*} - \frac{\left(k_p + sk_d\right)G_{LPF}(s)}{1+G_{f\,sec}(s)G_{PLL}(s)}P \qquad (8.6)$$

where $G_{LPF}(s)$, $G_{fsec}(s)$, and $G_{PLL}(s)$ are expressed as

$$G_{LPF}(s) = \frac{\omega_c}{s+\omega_c}, \quad G_{f\,sec}(s) = k_{pf} + \frac{k_{if}}{s}, \quad G_{PLL}(s) = \frac{1}{\tau s + 1} \qquad (8.7)$$

Figure 8.5 shows the operation principle of secondary control, which removes frequency and voltage amplitude deviations caused by the primary control loop. The characteristic of secondary control for the frequency restoration is shown in Fig. 8.5a. It can be seen that secondary control shifts up the primary response so that frequency reaches to the nominal value. As shown in Fig. 8.5a, the points of A and B are the nominal frequencies of the DG1 and DG2, respectively. The operation points of DG1 and DG2 deviate from the nominal frequencies and operate at the points of C and D when a transient increase of load is applied in the system. The idling frequency changes and the operation points of DG1 and DG2 shift to new operating points of C^{*} and D^{*} after the secondary controller is applied in the control system. Without this action, the frequency of the MG is load dependent.

2. *Voltage control*: A similar approach can be used as in the frequency secondary control, in which each DG unit measures the voltage error, and tries to compensate the voltage deviation caused by the Q-E droop. As shown in Fig. 8.5b, the secondary control is able to remove voltage deviations caused by primary control in DG unit and the voltage amplitude restoration can be achieved. Figure 8.6 shows the simplified control diagram in this case. The closed-loop voltage dynamic model can be obtained as

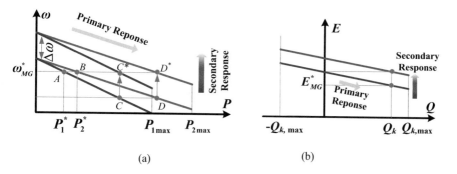

(a) (b)

Fig. 8.5 Secondary control versus primary control response. (**a**) Frequency restoration. (**b**) Voltage amplitude restoration

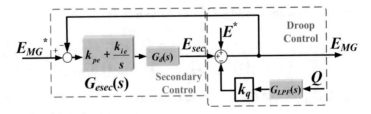

Fig. 8.6 Block diagram of voltage amplitude control for a DG unit

$$E_{MG} = \frac{G_{esec}(s)G_d(s)}{1+G_{esec}(s)}E_{MG}^* - \frac{k_q G_{LPF}(s)}{1+G_{esec}(s)}Q, \quad G_{esec}(s) = k_{pe} + \frac{k_{ie}}{s} \quad (8.8)$$

8.3   Modeling and Controller Design of Islanded Microgrid

8.3.1   Small-Signal Model of Power Controller

This section presents the small-signal model of the primary and secondary controllers with additional phase-shift loop, emphasizing stability of the MG power controller. The small-signal dynamics of the restoration control can be obtained by linearizing Eq. (8.4):

$$\Delta\omega_{sec} = -k_{pf}\Delta\omega_{MG} - \frac{k_{if}}{s}\Delta\omega_{MG}, \quad \Delta E_{sec} = -k_{pe}\Delta E_{MG} - \frac{k_{ie}}{s}\Delta E_{MG} \quad (8.9)$$

where the symbol Δ in Eq. (8.9) denotes the small deviation of the variable from the equilibrium point [2].

Taking Eqs. (8.1) to (8.5) into account, the droop control with centralized secondary restoration control can be obtained as

$$\omega = \omega_{sec} + \omega^* - k_p\left(P - P^*\right), \quad E_{MG} = E_{sec} + E_{MG}^* - k_q\left(Q - Q^*\right) \quad (8.10)$$

By linearizing Eq. (8.10), and substituting Eq. (8.9) for (8.10), the small-signal model can be written as

$$\begin{cases} \Delta\omega = -k_{pf}\Delta\omega_{MG} - \frac{k_{if}}{s}\Delta\omega_{MG} - k_p\Delta P \\ \\ \Delta E_{MG} = -k_{pe}\Delta E_{MG} - \frac{k_{ie}}{s}\Delta E_{MG} - k_q\Delta Q \end{cases} \quad (8.11)$$

The linearized small-signal models of P and Q can be written as

Fig. 8.7 Linearized model
of the additional phase-
shift loop for P-ω droop
control

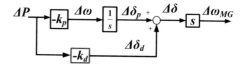

$$\begin{cases} \Delta \dot{P} = -\omega_c \Delta P + \omega_c \left(I_{o\alpha} \Delta v_{C\alpha} + I_{o\beta} \Delta v_{C\beta} + V_{C\alpha} \Delta i_{o\alpha} + V_{C\beta} \Delta i_{o\beta} \right) \\ \Delta \dot{Q} = -\omega_c \Delta Q + \omega_c \left(I_{o\alpha} \Delta v_{C\beta} - I_{o\beta} \Delta v_{C\alpha} + V_{C\beta} \Delta i_{o\alpha} - V_{C\alpha} \Delta i_{o\beta} \right) \end{cases}$$ (8.12)

Figure 8.7 shows the linearized model for frequency and power angle loop, and
the linearized inverter phase angle is:

$$\Delta \delta = \Delta \delta_p + \Delta \delta_d = \frac{1}{s} \Delta \omega - k_d \Delta P$$ (8.13)

Substituting $\Delta \omega$ of Eq. (8.11) for (8.13), we get

$$s \Delta \delta = -k_{pf} \Delta \omega_{MG} - \frac{k_{if}}{s} \Delta \omega_{MG} - k_p \Delta P - s k_d \Delta P$$ (8.14)

At this point, as shown in Fig. 8.7, note that the first derivative of the inverter
phase angle ($d\delta/dt$) is not the droop frequency (ω) but the MG frequency (ω_{MG}) and
the first state variables for the state equation model can be obtained as

$$\Delta \dot{\delta}(t) = \Delta \omega_{MG}(t) \quad \text{or} \quad s \Delta \delta(s) = \Delta \omega_{MG}(s)$$ (8.15)

Considering that the frequency is the first-order derivative of the phase
angle, we get

$$s \Delta \delta = s \Delta \delta_p + s \Delta \delta_d = \Delta \omega_{MG} = \Delta \omega + \Delta \omega_d$$ (8.16)

Based on Eqs. (8.11), (8.13), and (8.16), the equations that relate the frequency
shift and secondary control of the inverter can be calculated due to the active power
deviation from the equilibrium point, thus we get

$$\left(1 + k_{pf} \right) \Delta \dot{\omega}_{MG} = -k_{if} \Delta \omega_{MG} - \left(k_p + k_d \right) \Delta \dot{P}$$ (8.17)

Finally, according to ΔE_{MG} of Eq. (8.11), the derivative of E_{MG} can be obtained as

$$\left(1 + k_{pe} \right) \Delta \dot{E}_{MG} = -k_{ie} \Delta E_{MG} - k_q \Delta \dot{Q}$$ (8.18)

By rearranging Eqs. (8.9) to (8.18), the small-signal power controller model can
be written in a state-space form as in Eq. (8.19), which describes the behavior of the

states $\Delta\delta$, $\Delta\omega_{MG}$, ΔE_{MG} and ΔP, and ΔQ on the kth ($k = 1, 2$) inverter in function of the deviations of the active power and reactive power from the equilibrium point.

$$
\begin{bmatrix} \dot{\Delta\delta_k} \\ \Delta\omega_{MGk} \\ \Delta E_{MGk} \\ \Delta P_k \\ \Delta Q_k \end{bmatrix} = \mathbf{M_k} \begin{bmatrix} \Delta\delta_k \\ \Delta\omega_{MGk} \\ \Delta E_{MGk} \\ \Delta P_k \\ \Delta Q_k \end{bmatrix} + \mathbf{N_k} \begin{bmatrix} \Delta v_{C\alpha\beta k} \\ \Delta i_{o\alpha\beta k} \end{bmatrix}
\tag{8.19}
$$

Or symbolically, represented as

$$
\left[\Delta \dot{X}_k \right] = \mathbf{M_k} \left[\Delta X_k \right] + \mathbf{N_k} \left[\Delta S_k \right]
\tag{8.20}
$$

where $\mathbf{M_k}$ and $\mathbf{N_k}$ are derived as

$$
\mathbf{M_k} = \begin{bmatrix}
0 & 1 & 0 & 0 & 0 \\
0 & \dfrac{-k_{if}}{1+k_{pf}} & 0 & 0 & 0 \\
0 & 0 & \dfrac{-k_{ie}}{1+k_{pe}} & 0 & \dfrac{-k_q\omega_c}{1+k_{pe}} \\
0 & 0 & 0 & -\omega_c & 0 \\
0 & 0 & 0 & 0 & -\omega_c
\end{bmatrix}, \quad
\mathbf{N_k} = \begin{bmatrix}
0 & 0 & 0 & 0 \\
\eta I_{o\alpha k} & \eta I_{o\beta k} & \eta V_{C\alpha k} & \eta V_{C\beta k} \\
-\gamma I_{o\alpha k} & \gamma I_{o\beta k} & -\gamma V_{C\alpha k} & \gamma V_{C\beta k} \\
\omega_c I_{o\alpha k} & \omega_c I_{o\beta k} & \omega_c V_{C\alpha k} & \omega_c V_{C\beta k} \\
\omega_c I_{o\alpha k} & -\omega_c I_{o\beta k} & \omega_c V_{C\beta k} & -\omega_c V_{C\alpha k}
\end{bmatrix}
\tag{8.21}
$$

with

$$
\eta = \frac{-\left(k_p + k_d\right)\omega_c}{1+k_{pf}}, \quad \gamma = \frac{k_q\omega_c}{1+k_{pe}}.
\tag{8.22}
$$

Taking the Laplace transformation on both the sides of (8.20), using initial conditions of $x_{init} = 0$, we get

$$
s\left[\Delta X_k\left(s\right) \right] = \mathbf{M_k}\left[\Delta X_k\left(s\right) \right] + \mathbf{N_k}\left[\Delta S_k\left(s\right) \right]
\tag{8.23}
$$

$$
\left(s\mathbf{I}_{5\times5} - \mathbf{M_k}\right)\left[\Delta X_k\left(s\right) \right] = \mathbf{N_k}\left[\Delta S_k\left(s\right) \right]
\tag{8.24}
$$

where $\mathbf{I}_{5\times5}$ is a fifth order identity matrix. Then, assuming $(s\mathbf{I}_{5\times5}\text{-}\mathbf{M_k})$ is nonsingular, $\Delta X_k(s)$ can be calculated as

$$
\left[\Delta X_k\left(s\right) \right] = \left(s\mathbf{I}_{5\times5} - \mathbf{M_k}\right)^{-1} \mathbf{N_k}\left[\Delta S_k\left(s\right) \right]
\tag{8.25}
$$

By using adjoint matrix $adj(s\mathbf{I}_{5\times5}\text{-}\mathbf{M_k})$, $\Delta X_k(s)$ can be rewritten as

Table 8.1 The parameters of the primary and secondary controllers

Symbol	Parameter	Value
ω_c	Measuring filter cut-off frequency	10π rad/s
k_p, k_q	Frequency and voltage droop coefficient	0.0001 rad/s/W, 0.0001 V/Var
k_{pf}, k_{if}	Frequency proportional and integral term of the secondary compensator	0.8, 10 s^{-1}
k_{pe}, k_{ie}	Amplitude proportional and integral term of the secondary compensator	0.8, 10 s^{-1}
k_d	Additional phase-shift coefficient	0.000005 rad/W
τ	PLL time constant	50 ms
R- $v_{,}f, R_{v,5}, R_{v,7}, R_{v,11},$ and $R_{v,13}$	Virtual resistances	6, 1, 1, 1, and 1 Ω
L+ $v_{,}f, L_{v,5}, L_{v,7}, L_{v,11},$ and $L_{v,13}$	Virtual inductances	6, 2, 2, 1.5 and 1.5 mH

$$\left[\Delta X_k\left(s\right)\right] = \frac{adj\left(s\mathbf{I}_{5\times5} - \mathbf{M_k}\right)\mathbf{N_k}\left[\Delta S_k\left(s\right)\right]}{\left|\left(s\mathbf{I}_{5\times5} - \mathbf{M_k}\right)\right|} \tag{8.26}$$

To ensure stable system, the poles of the denominator of Eq. (8.26) must lie in the left-hand side of the s-plane, thus

$$D(s) = \left|\left(s\mathbf{I}_{5\times5} - \mathbf{M_k}\right)\right| = 0 \tag{8.27}$$

Substituting parameters of Table 8.1 to Eq. (8.27), the eigenvalues of the matrix $\mathbf{M_k}$ defined by Eq. (8.21) are calculated as

$$\lambda_1 = 0, \quad \lambda_2 = \lambda_3 = -5.5556, \quad \lambda_4 = \lambda_5 = -31.4159. \tag{8.28}$$

Note that all the nonzero poles of the matrix $\mathbf{M_k}$ are real, and the system is over-damped. According to Ref. [2], the calculation of δ derivative presents a high variation level due to the active power ripple, especially under nonlinear load conditions. ω instead of ω_{MG} is used as the frequency feedback for the secondary controller in the experiment.

8.3.2   Virtual Impedance Loop

Virtual resistance enhances system damping without additional power loss, since it is provided by a control loop and it is possible to implement it without decreasing system efficiency. When virtual inductance is utilized, the DG output impedance becomes more inductive, decreasing P and Q coupling, enhancing the system stability, and reducing power oscillations and circulating currents.

As shown in Fig. 8.3, the voltage drop across the virtual positive- and negative-sequence impedance, and the virtual variable harmonic impedance loops in $\alpha\beta$ reference frame are derived as

$$\begin{cases} v_{v\alpha,f}^{+}\left(s\right)=R_{v,f}^{+}i_{o\alpha,f}^{+}-\omega_{0}L_{v,f}^{+}i_{o\beta,f}^{+} \\ v_{v\beta,f}^{+}\left(s\right)=R_{v,f}^{+}i_{o\beta,f}^{+}+\omega_{0}L_{v,f}^{+}i_{o\alpha,f}^{+} \end{cases} \tag{8.29}$$

$$\begin{cases} v_{v\alpha,f}^{-}\left(s\right)=R_{v,f}^{-}i_{o\alpha,f}^{-}+\omega_{0}L_{v,f}^{-}i_{o\beta,f}^{-} \\ v_{v\beta,f}^{-}\left(s\right)=R_{v,f}^{-}i_{o\beta,f}^{-}-\omega_{0}L_{v,f}^{-}i_{o\alpha,f}^{-} \end{cases} \tag{8.30}$$

$$\begin{cases} v_{v\alpha,h}\left(s\right)=R_{v,h}i_{o\alpha,h}+h\omega_{0}L_{v,h}i_{o\beta,h} \\ v_{v\beta,h}\left(s\right)=R_{v,h}i_{o\beta,h}-h\omega_{0}L_{v,h}i_{o\alpha,h} \end{cases} \tag{8.31}$$

where $R+$ v,f and $L+$ v,f are the virtual fundamental frequency positive sequence resistance and inductance, $R-$ v,f and $L-$ v,f represent the virtual fundamental frequency negative sequence resistance and inductance, and h denotes the dominant harmonic components, which are $-5, 7, -11, 13$, etc., and ω_0 represents the system fundamental frequency.

At the fundamental frequency, the virtual positive sequence impedance loop is designed to be mainly inductive to improve the reactive power sharing based on the Q-E droop, and the problem of the presence of high R/X ratio which causes a coupling in the control of active and reactive power when using the conventional droop controllers has been resolved. The virtual negative sequence impedance is designed to be resistive to minimize the negative sequence circulating current among the DGs. The size of negative inductance needs to be kept smaller than the effective inductance to guarantee the stability of the virtual variable harmonic impedance loop at harmonic frequencies, and the larger the positive resistance in the virtual variable harmonic impedance loop, the better the sharing of harmonic power can be achieved.

In order to extract the fundamental positive sequence and negative sequence currents as well as the dominant harmonic currents, a set of Park transformation and the moving average filters are presented for realizing the load current decomposition, which is shown in Fig. 8.8a. The moving average filters are linear-phase finite impulse response filters that are easy to realize in practice, are cost effective in terms of the computational burden, and can act as ideal low-pass filters if certain conditions hold [13, 14]. The transfer function of the moving average filter can be simply presented as

$$G_{MAF}\left(s\right)=\frac{1-e^{-T_{\omega}s}}{T_{\omega}s} \tag{8.32}$$

where T_{ω} is referred to as the window length. The moving average filter passes the dc component, and completely blocks the frequency components of integer multiples of $1/T_{\omega}$ in hertz.

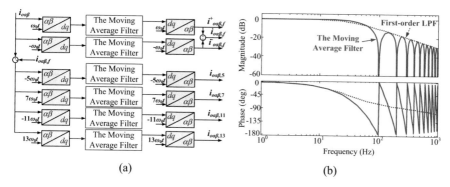

(a)   (b)

Fig. 8.8 The proposed moving average filter-based sequence decomposition of fundamental positive- and negative-sequence, and harmonic components. (**a**) Block diagram. (**b**) Bode plots for the moving average filter

To provide a means of comparison, the transfer function of the first-order counterpart of the moving average filter is obtained as Eq. (8.33) by approximating the delay term in Eq. (8.32) by the first-order Padé approximation.

$$G_{MAF}(s)\Big|_{e^{-T_\omega s} \approx \frac{1-T_\omega s/2}{1+T_\omega s/2}} \approx \frac{1}{T_\omega s/2 + 1} \qquad (8.33)$$

The moving average filter with 0.01 s window length (T_ω) is used in the sequence decomposition and the bode plots of the moving average filter and the first-order low-pass filter (LPF) are shown in Fig. 8.8b. It can be observed that the moving average filter results in notches at the concerned harmonic frequencies. Hence the accuracy of the sequence decomposition is significantly improved by using the moving average filter.

8.3.3   Inner Voltage and Current Control Loops

From system control diagram of Fig. 8.3, the simplified model of the inner loops is derived, as shown in Fig. 8.9a. In order to overcome the drawbacks of the passive damping method, the active damping of a virtual resistor in parallel with the capacitor is used to avoid resonance and enhance stability of the inner loop controller.

The voltage loop reference signals are modified by the virtual impedance loop which contains the virtual positive- and negative-sequence impedance, and the virtual variable harmonic impedance loops are shown in Fig. 8.9a. Then, the output voltage of the DG unit can be derived as

$$v_{C\alpha\beta}(s) = G(s)v_{\alpha\beta}^* - \left(G(s)Z_{v\alpha\beta}(s) + Z_{o\alpha\beta}(s)\right)i_{o\alpha\beta} \qquad (8.34)$$

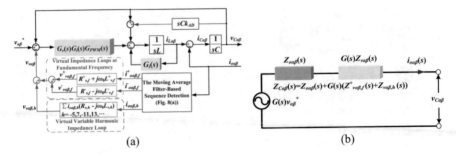

Fig. 8.9 Voltage and current loops integrated with virtual impedance loop. (**a**) Simplified model of the inner loops. (**b**) Equivalent impedance model of a VSI

where $G(s)$, $Z_{v\alpha\beta}$, and $Z_{o\alpha\beta}$ are the closed-loop voltage transfer function, resistive-inductive virtual impedance, and the output impedance without virtual impedance loops, respectively. The virtual positive sequence impedance loop at fundamental frequency is only for attenuating circulating current, which can be omitted. The transfer functions in Eq. (8.34) are derived as [13, 14]:

$$G(s) = \frac{G_v(s)G_i(s)G_{PWM}(s)}{LCs^2 + (Cs + G_v(s))G_i(s)G_{PWM}(s) + k_{AD}Cs + 1} \quad (8.35)$$

$$Z_{v\alpha\beta}(s) = Z_{v\alpha\beta,f}^+(s) + Z_{v\alpha\beta,h}(s)$$
$$= \begin{bmatrix} R_{v,f}^+ & -\omega_0 L_{v,f}^+ \\ \omega_0 L_{v,f}^+ & R_{v,f}^+ \end{bmatrix} + \begin{bmatrix} R_{v,h} & h\omega_0 L_{v,h} \\ -h\omega_0 L_{v,h} & R_{v,h} \end{bmatrix} \quad (8.36)$$

$$Z_{o\alpha\beta}(s) = \frac{Ls + G_i(s)}{LCs^2 + (Cs + G_v(s) + k_{AD}Cs)G_i(s)G_{PWM}(s) + 1} \quad (8.37)$$

Under nonlinear load conditions, the dominant harmonic components should be taken into consideration for the voltage and current controllers in order to suppress output voltage harmonics. The transfer function of the voltage and current controllers are

$$G_v(s) = k_{pv} + \frac{k_{rv}s}{s^2 + \omega_0^2} + \sum_{h=-5,7,-11,13} \frac{k_{hv}s}{s^2 + (\omega_0 h)^2} \quad (8.38)$$

$$G_i(s) = k_{pi} + \frac{k_{ri}s}{s^2 + \omega_0^2} + \sum_{h=-5,7,-11,13} \frac{k_{hi}s}{s^2 + (\omega_0 h)^2} \quad (8.39)$$

where k_{pv} and k_{pi} are the proportional coefficients, k_{rv} and k_{ri} are the resonant coefficients at the fundamental frequency, k_{hv} and k_{hi} represent the voltage and current resonant controller coefficients for the hth order harmonic component.

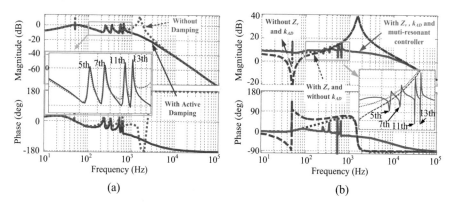

Fig. 8.10 Bode plots of the closed-loop voltage gain and the output impedance of a DG unit. (**a**) Closed-loop voltage gain of $G(s)$ with PR plus multiresonant controller. (**b**) The output impedance with PR and multiresonant controllers

The total output impedance with the virtual impedance loop can be derived as

$$Z_{C\alpha\beta}(s) = G(s)\left(Z_{v\alpha\beta,f}^{+}(s) + Z_{v\alpha\beta,h}(s)\right) + Z_{o\alpha\beta}(s) \tag{8.40}$$

From Eqs. (8.35) to (8.40), the equivalent impedance model of a DG unit can be derived, as shown in Fig. 8.9b.

By using the closed-loop model described by Eqs. (8.35) to (8.40), the bode plots of the closed-loop voltage gain and the output impedance with and without AD method are illustrated in Fig. 8.10. From Fig. 8.10a, it can be observed that the AD method ensures effective damping at LCL resonance frequency and the system shows sufficient stability margin. The gains of the closed-loop voltage controller are unity at the fundamental and 5th, 7th, 11th, and 13th harmonic frequencies, respectively, which means the system obtains the zero-error tracking capability at both the fundamental frequency and target characteristic harmonic frequencies.

The total output impedance $Z_{C\alpha\beta}(s)$ of a DG unit under PR controller plus virtual impedance loops which contain the virtual positive- and negative-sequence impedance, and the virtual variable harmonic impedance loops without AD are given in Fig. 8.10b. It shows that the total output impedance is about 40 dB without AD under resonance frequency of the LC filter and about 20 dB under the considered frequencies such as $-5f_0$, $7f_0$, $-11f_0$, $13f_0$. Large output impedance leads to a large harmonic voltage drop under current harmonics, which distorts the output voltage. The total output impedance of the DG under PR plus multiresonant controller with virtual impedance loops using AD method, $Z_{C\alpha\beta}(s)$ is greatly reduced, and lower voltage distortion can be expected under nonlinear loads.

8.4   Simulation Results Using EMTP

This section presents the simulation results of islanded microgrid obtained from EMTP, where four DG units are connected to the common AC bus through feeder impedances, and an abrupt increase of resistive load is applied to test the dynamic response of the microgrid system. The nominal parameters of the LC filter for each DG are $L_1 = 1.8$ mH, $C = 25$ μF, and the nominal feeder impedance is $L_2 = 1.8$ mH with parasitic resistance of 0.5 Ω. Similar to the case study in the previous section, the control algorithms are implemented in the stationary frame alpha-beta frame, hence a factor of 1.5 should be multiplied in the instantaneous active and reactive powers when three-phase quantities are evaluated in the following subsections. Besides, due to the space limitations, the case of identical power-stage parameters is omitted in the forthcoming subsections.

8.4.1   Conventional Droop Control with Unequal Feeder Impedance

In the first case, the results under convention droop control are given, for the sake of comparison. The nominal load 2 Ω is applied, with a transient increase of another 2 Ω resistance to test the dynamic features of the closed-loop control system. The inner loop controller design procedure is consistent with the case in the previous chapters, hence would be omitted here. Next, the microgrid under nonidentical feeder inductance scenario is first simulated using EMTP software.

1. Nonidentical Feeder Inductance Scenario

Figure 8.11 shows the simulation results of the islanded microgrid under non-identical feeder inductance scenario with droop coefficient $k_p = k_q = 0.0001$, and the line inductances of DG2, DG3, DG4 are -0.1 pu, 0.1 pu, 0.2 pu perturbed from their nominal values. Figure 8.11a, b shows the reference voltage amplitude and the reference frequency of each DG, respectively. Figure 8.11c, d shows the instantaneous active and reactive powers of each DG unit, and the damping control output of each DG, respectively. Figure 8.12 shows the similar case as in Fig. 8.11, with the droop coefficient $k_p = k_q = 0.0005$. Figures 8.13 shows the similar case as in Fig. 8.11, with the droop coefficient $k_p = k_q = 0.0008$.

In these cases, the closed-loop control is enabled at $t = 0.02$ s, and a transient increase of load is applied at $t = 0.3$ s, and tripped at $t = 1.0$ s, to check the dynamic performance of the islanded microgrid. In the forthcoming subsections, the load disturbances are consistent with this case, hence would not be repeatedly described for the sake of brevity.

As expected, a higher droop gain would result in a higher reference voltage amplitude and frequency dip, and active and reactive power sharing are achieved under this case. However, an excessive high droop gain would cause oscillating

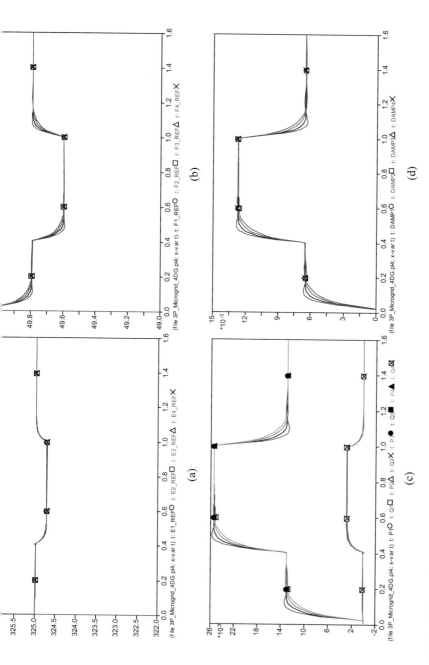

Fig. 8.11 Simulation results under nonidentical feeder inductance scenario with droop coefficient $k_p = k_q = 0.0001$ (the line inductances of DG2, DG3, DG4 are -0.1 pu, 0.1 pu, 0.2 pu perturbed from nominal values). (**a**) Reference voltage amplitude of each DG; (**b**) reference frequency of each DG; (**c**) instantaneous active and reactive powers of each DG; (**d**) damping control output of each DG

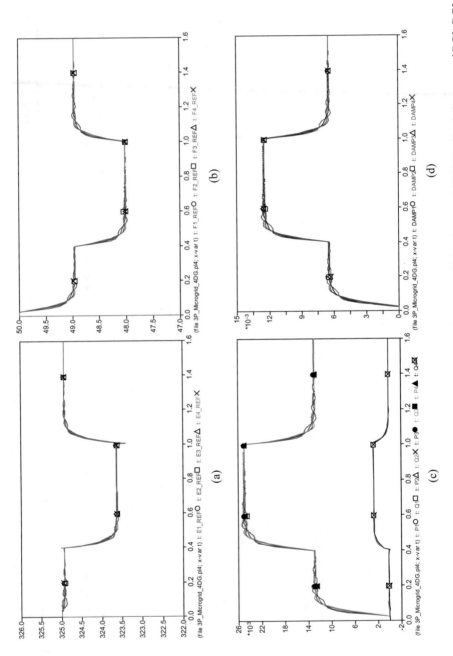

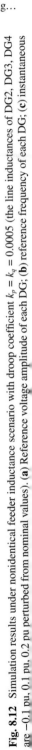

Fig. 8.12 Simulation results under nonidentical feeder inductance scenario with droop coefficient $k_p = k_q = 0.0005$ (the line inductances of DG2, DG3, DG4 are −0.1 pu, 0.1 pu, 0.2 pu perturbed from nominal values). (**a**) Reference voltage amplitude of each DG; (**b**) reference frequency of each DG; (**c**) instantaneous

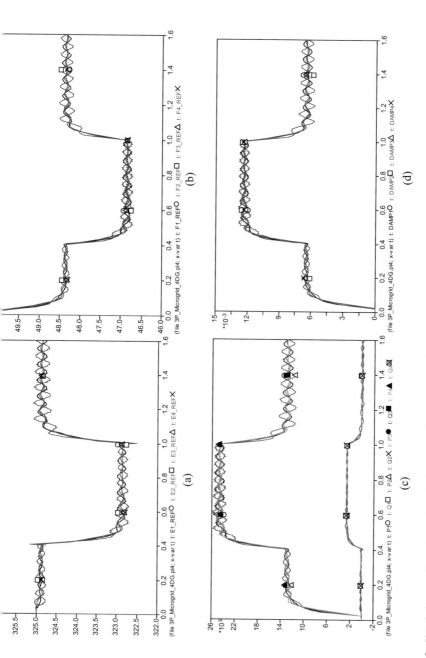

Fig. 8.13 Simulation results under nonidentical feeder inductance scenario with droop coefficient $k_p = k_q = 0.0008$ (the line inductances of DG2, DG3, DG4 are -0.1 pu, 0.1 pu, 0.2 pu perturbed from nominal values). (**a**) Reference voltage amplitude of each DG; (**b**) reference frequency of each DG; (**c**) instantaneous active and reactive powers of each DG; (**d**) damping control output of each DG

behavior or system instability. The additional power loop damping term provides auxiliary damping to the closed-loop control, which would be helpful to stabilize the system, yet the parameter selection is crucial to achieve an optimized performance in terms of the steady-state accuracy and dynamic response.

2. Nonidentical Feeder Inductance and Resistance Scenario

Figure 8.14 shows the simulation results of the islanded microgrid under nonidentical feeder inductance and resistance scenario with droop coefficient $k_p = k_q = 0.0001$, and the line inductances of DG2, DG3, DG4 are −0.1 pu, 0.1 pu, 0.2 pu perturbed from their nominal values, and the line resistances of DG2, DG3, DG4 are 0.2 pu, 0.4 pu, 0.6 pu perturbed from nominal values. Figure 8.15 shows the similar case as in Fig. 8.14, with the droop coefficient $k_p = k_q = 0.0005$. Figure 8.16 shows the similar case as in Fig. 8.14, with the droop coefficient $k_p = k_q = 0.0008$.

Similar dynamic response can be observed from Figs. 8.14 to 8.16, as compared to the case studies in Figs. 8.11, 8.12, and 8.13. In addition, nonidentical reference voltage amplitude and reactive power sharing can be observed due to the difference in the feeder resistance. The active power sharing, as shown in Figs. 8.14, 8.15, and 8.16, is also ensured due to the inherent power sharing capabilities of the droop control. Besides, a large droop gain would cause oscillating behavior in the reference voltage amplitude, reference frequency, and instantaneous active and reactive powers. A compromise should be achieved between the voltage amplitude and frequency regulation, and the instantaneous active and reactive power sharing accuracy. Hence, in order to realize voltage amplitude and frequency restoration, the secondary control would be utilized, which would be discussed in the following subsection.

8.4.2 Proposed Secondary Control Scheme with Unequal Feeder Impedance

Next, the effect of secondary control is simulated under nonidentical feeder inductance and resistance scenario, and the influence of low-bandwidth communication (LBC) is also evaluated on the power sharing performance and dynamic response of the islanded AC microgrid.

1. Unequal Feeder Inductance and Resistance Scenario without LBC Delay

Figure 8.17 shows the simulation results of the islanded microgrid under nonidentical feeder inductance and resistance scenario without LBC delay when the droop coefficient $k_p = k_q = 0.0001$, and the line inductances of DG2, DG3, DG4 are −0.1 pu, 0.1 pu, 0.2 pu perturbed from their nominal values, and the line resistances of DG2, DG3, DG4 are 0.2 pu, 0.4 pu, 0.6 pu perturbed from nominal values. Figure 8.18 shows the similar case as in Fig. 8.17, with the droop coefficient

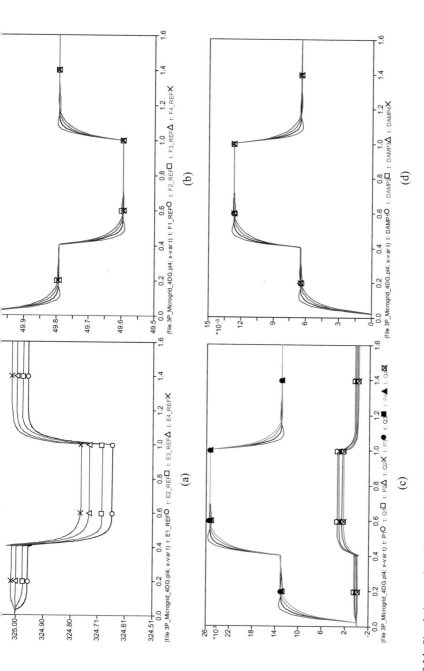

Fig. 8.14 Simulation results under nonidentical feeder inductance and resistance scenario with droop coefficient $k_p = k_q = 0.0001$ (the line inductances of DG2, DG3, DG4 are -0.1 pu, 0.1 pu, 0.2 pu perturbed from nominal values, and the line resistances of DG2, DG3, DG4 are 0.2 pu, 0.4 pu, 0.6 pu perturbed from nominal values). (**a**) Reference voltage amplitude of each DG; (**b**) reference frequency of each DG; (**c**) instantaneous active and reactive powers of each DG; (**d**) damping control output of each DG

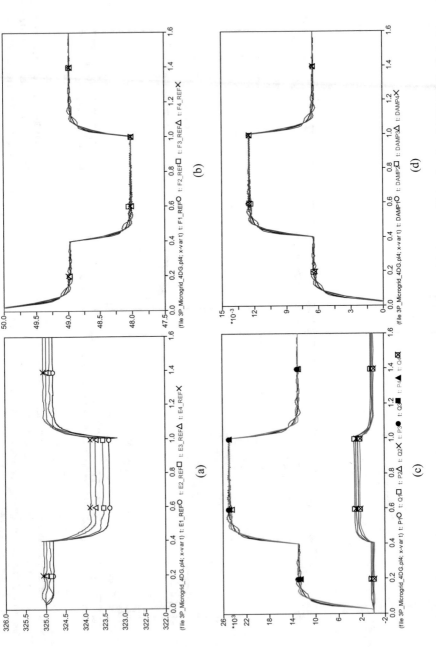

Fig. 8.15 Simulation results under nonidentical feeder inductance and resistance scenario with droop coefficient $k_p = k_q = 0.0005$ (the line inductances of DG2, DG3, DG4 are −0.1 pu, 0.1 pu, 0.2 pu perturbed from nominal values, and the line resistances of DG2, DG3, DG4 are 0.2 pu, 0.4 pu, 0.6 pu perturbed from

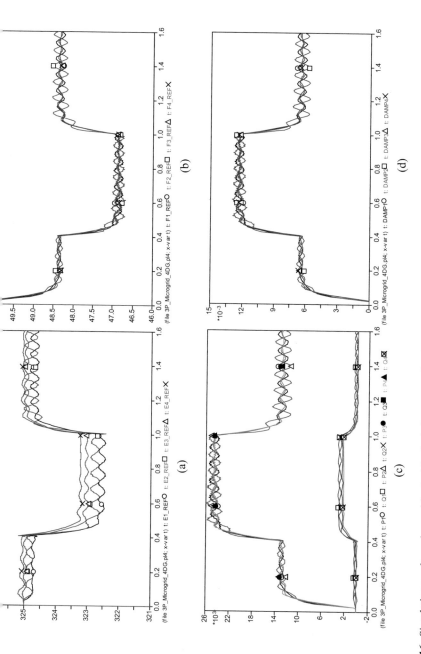

Fig. 8.16 Simulation results under nonidentical feeder inductance and resistance scenario with droop coefficient $k_p = k_q = 0.0008$ (the line inductances of DG2, DG3, DG4 are −0.1 pu, 0.1 pu, 0.2 pu perturbed from nominal values, and the line resistances of DG2, DG3, DG4 are 0.2 pu, 0.4 pu, 0.6 pu perturbed from nominal values). (**a**) Reference voltage amplitude of each DG; (**b**) reference frequency of each DG; (**c**) instantaneous active and reactive powers of each DG; (**d**) damping control output of each DG

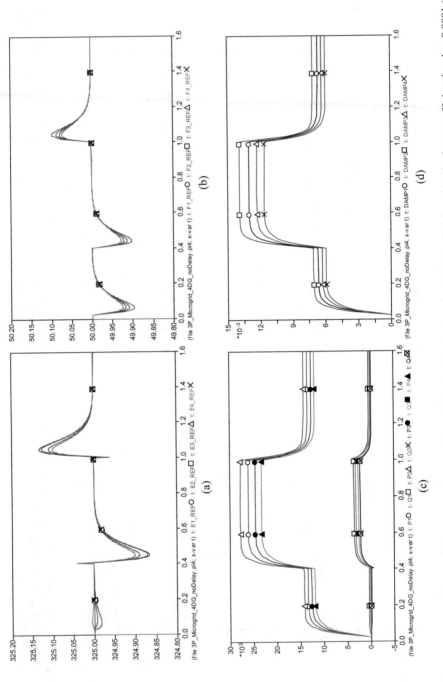

Fig. 8.17 Simulation results under nonidentical feeder inductance and resistance scenario without LBC delay when the droop coefficient $k_p = k_q = 0.0001$ (the line inductances of DG2, DG3, DG4 are -0.1 pu, 0.1 pu, 0.2 pu perturbed from nominal values, and the line resistances of DG2, DG3, DG4 are 0.2 pu, 0.4 pu,

$k_p = k_q = 0.0005$. Figure 8.19 shows the similar case as in Fig. 8.17, with the droop coefficient $k_p = k_q = 0.0008$.

Figures 8.17, 8.18, and 8.19 show that a higher droop gain would result in a higher transient overshoot in the reference voltage amplitude and frequency, yet the active power sharing accuracy is improved. Due to the effect of the secondary control loop, the voltage amplitude and frequency restoration are effectively achieved in the steady-state conditions.

Figure 8.20 shows the simulation results under nonidentical feeder inductance and resistance scenario without LBC delay when the droop coefficient $k_p = k_q = 0.0005$, and secondary controller proportional gain $k_{pf} = k_{pe} = 0.5$. The line inductances of DG2, DG3, DG4 are −0.1 pu, 0.1 pu, 0.2 pu perturbed from nominal values, and the line resistances of DG2, DG3, DG4 are 0.2 pu, 0.4 pu, 0.6 pu perturbed from nominal values, respectively. Figure 8.21 shows the case when secondary controller proportional gain $k_{pf} = k_{pe} = 1.0$. As expected, an increased secondary controller gain would enhance the dynamic response of the system, with a reduced overshoot in the reference voltage amplitude and frequency, as shown in Figs. 8.20 and 8.21.

2. Nonidentical Feeder Inductance and Resistance Scenario with LBC Delay

Next, the effect of the low-bandwidth communication (LBC) is evaluated under nonidentical feeder inductance and resistance scenario. Two delay models, namely, the low-pass filter model, and the lead-lag compensator model, are tested and compared in terms of dynamic response and power sharing performance, as discussed in the following subsections.

(a) Delay Model 1: Compare the Effect of LBC Delay

Firstly, the control delay is approximated using the low-pass filter (LPF). Figure 8.22 shows the simulation results under the nonidentical feeder inductance and resistance scenario with the LBC delay $T_d = 5$ ms, when the droop coefficient $k_p = k_q = 0.0005$, secondary controller proportional gain $k_{pf} = k_{pe} = 0.1$ The line inductances of DG2, DG3, DG4 are −0.1 pu, 0.1 pu, 0.2 pu perturbed from the nominal values, and the line resistances of DG2, DG3, DG4 are 0.2 pu, 0.4 pu, 0.6 pu perturbed from the nominal values. Figure 8.23 shows the similar case as in Fig. 8.22, with the LBC delay $T_d = 50$ ms. In addition, Fig. 8.24 shows the similar case as in Fig. 8.22, with the LBC delay $T_d = 100$ ms. It shows in Figs. 8.22 to 8.24 that a larger LBC delay results in a higher transient overshoot in the reference voltage amplitude and frequencyand a sluggish dynamic response. Moreover, the steady-state power sharing performance is insensitive to the LBC delay in this scenario.

(b)Delay Model 2: Compare the Effect of LBC Delay

Next, the control delay is approximated using the lead-lag compensator model. Figure 8.25 shows the simulation results under the nonidentical feeder inductance and resistance scenario with the LBC delay $T_d = 5$ ms, when the droop coefficient $k_p = k_q = 0.0005$, secondary controller proportional gain $k_{pf} = k_{pe} = 0.1$. The line

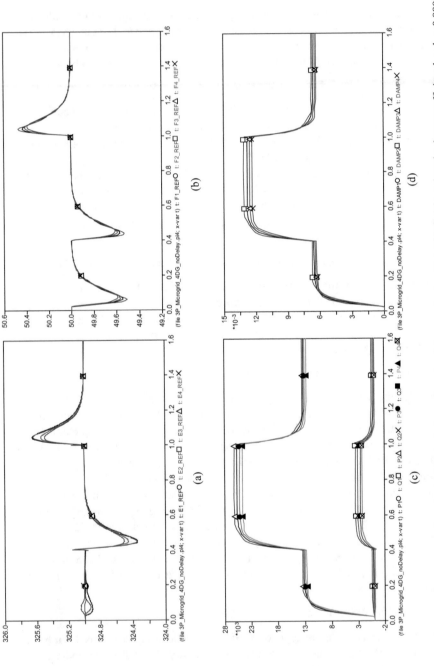

Fig. 8.18 Simulation results under nonidentical feeder inductance and resistance scenario without LBC delay when the droop coefficient $k_p = k_q = 0.0005$ (the line inductances of DG2, DG3, DG4 are −0.1 pu, 0.1 pu, 0.2 pu perturbed from nominal values, and the line resistances of DG2, DG3, DG4 are 0.2 pu, 0.4 pu,

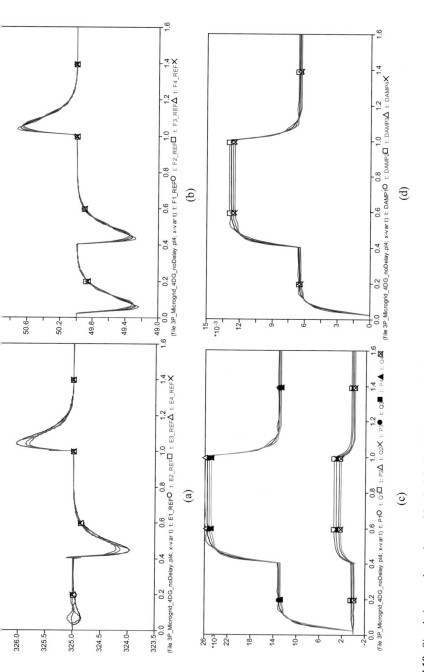

Fig. 8.19 Simulation results under nonidentical feeder inductance and resistance scenario without LBC delay when the droop coefficient $k_p = k_q = 0.0008$ (the line inductances of DG2, DG3, DG4 are -0.1 pu, 0.1 pu, 0.2 pu perturbed from nominal values, and the line resistances of DG2, DG3, DG4 are 0.2 pu, 0.4 pu, 0.6 pu perturbed from nominal values). (**a**) Reference voltage amplitude of each DG; (**b**) reference frequency of each DG; (**c**) instantaneous active and reactive powers of each DG; (**d**) damping control output of each DG

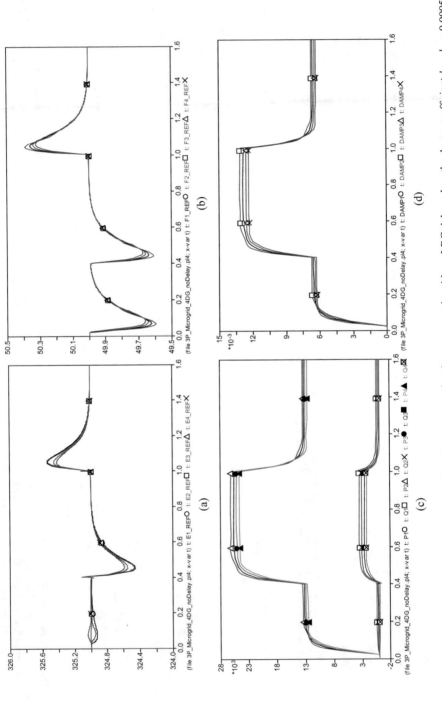

Fig. 8.20 Simulation results under nonidentical feeder inductance and resistance scenario without LBC delay when the droop coefficient $k_p = k_q = 0.0005$, secondary controller proportional gain $k_{pf} = k_{pe} = 0.5$ (the line inductances of DG2, DG3, DG4 are -0.1 pu, 0.1 pu, 0.2 pu perturbed from nominal values, and

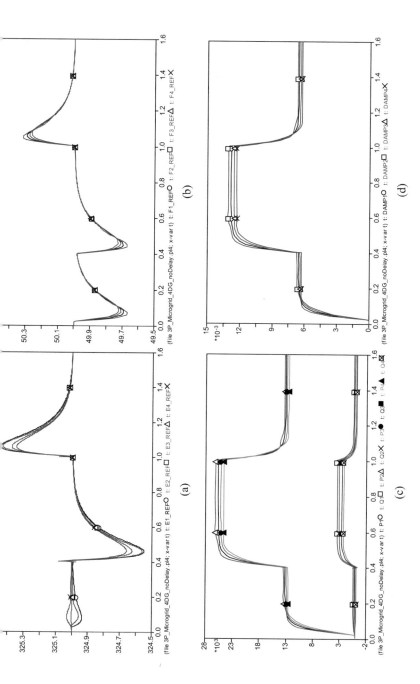

Fig. 8.21 Simulation results under nonidentical feeder inductance and resistance scenario without LBC delay when the droop coefficient $k_p = k_q = 0.0005$, secondary controller proportional gain $k_{pf} = k_{pe} = 1.0$ (the line inductances of DG2, DG3, DG4 are −0.1 pu, 0.1 pu, 0.2 pu perturbed from nominal values, and the line resistances of DG2, DG3, DG4 are 0.2 pu, 0.4 pu, 0.6 pu perturbed from nominal values). (**a**) Reference voltage amplitude of each DG; (**b**) reference frequency of each DG; (**c**) instantaneous active and reactive powers of each DG; (**d**) damping control output of each DG

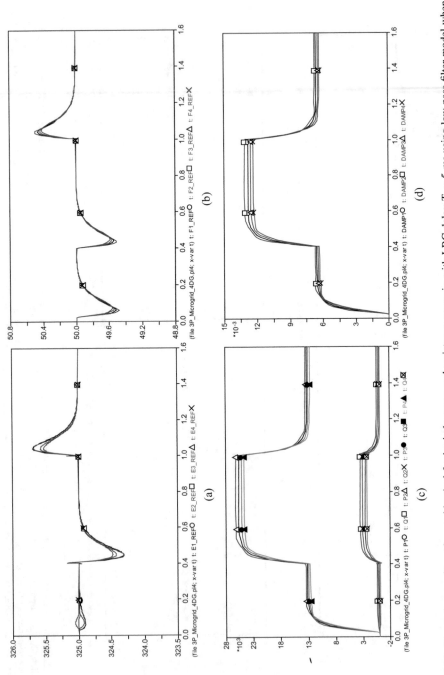

Fig. 8.22 Simulation results under nonidentical feeder inductance and resistance scenario with LBC delay $T_d = 5$ ms using low-pass filter model when the droop coefficient $k_p = k_q = 0.0005$, secondary controller proportional gain $k_{pf} = k_{pe} = 0.1$ (the line inductances of DG2, DG3, DG4 are -0.1 pu, 0.1 pu, 0.2 pu

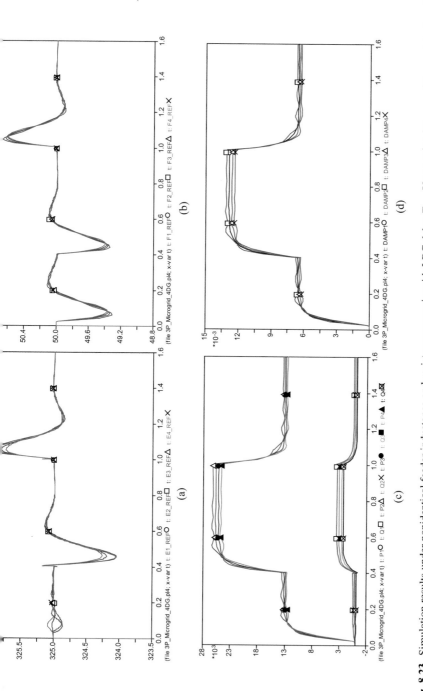

Fig. 8.23 Simulation results under nonidentical feeder inductance and resistance scenario with LBC delay $T_d = 50$ ms using low-pass filter model when the droop coefficient $k_p = k_q = 0.0005$, secondary controller proportional gain $k_{pf} = k_{pe} = 0.1$ (the line inductances of DG2, DG3, DG4 are -0.1 pu, 0.1 pu, 0.2 pu perturbed from nominal values, and the line resistances of DG2, DG3, DG4 are 0.2 pu, 0.4 pu, 0.6 pu perturbed from nominal values). (**a**) Reference voltage amplitude of each DG; (**b**) reference frequency of each DG; (**c**) instantaneous active and reactive powers of each DG; (**d**) damping control output of each DG

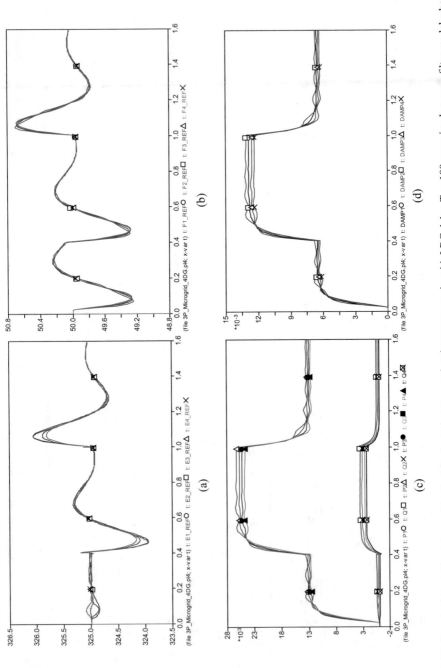

Fig. 8.24 Simulation results under nonidentical feeder inductance and resistance scenario with LBC delay $T_d = 100$ ms using low-pass filter model when the droop coefficient $k_p = k_q = 0.0005$, secondary controller proportional gain $k_{pf} = k_{pe} = 0.1$ (the line inductances of DG2, DG3, DG4 are −0.1 pu, 0.1 pu, 0.2 pu perturbed from nominal values, and the line resistances of DG2, DG3, DG4 are 0.2 pu, 0.4 pu, 0.6 pu perturbed from nominal values). (a) Reference voltage

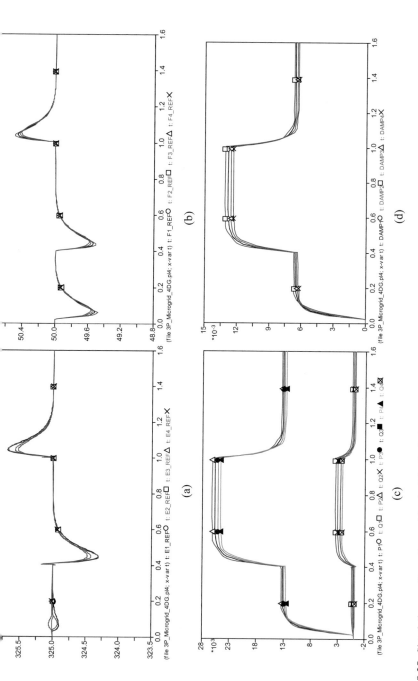

Fig. 8.25 Simulation results under nonidentical feeder inductance and resistance scenario with LBC delay $T_d = 5$ ms denoted using lead-lag compensator model when the droop coefficient $k_p = k_q = 0.0005$, secondary controller proportional gain $k_{pf} = k_{pe} = 0.1$ (the line inductances of DG2, DG3, DG4 are -0.1 pu, 0.1 pu, 0.2 pu perturbed from nominal values, and the line resistances of DG2, DG3, DG4 are 0.2 pu, 0.4 pu, 0.6 pu perturbed from nominal values). (**a**) Reference voltage amplitude of each DG; (**b**) reference frequency of each DG; (**c**) instantaneous active and reactive powers of each DG; (**d**) damping control output of each DG

inductances of DG2, DG3, DG4 are −0.1 pu, 0.1 pu, 0.2 pu perturbed from the nominal values, and the line resistances of DG2, DG3, DG4 are 0.2 pu, 0.4 pu, 0.6 pu perturbed from the nominal values. Figure 8.26 shows the similar case as in Fig. 8.25, with the LBC delay $T_d = 50$ ms. In addition, Fig. 8.27 shows the similar case as in Fig. 8.25, with the LBC delay $T_d = 60$ ms.

Figures 8.25, 8.26, and 8.27 show that similar steady-state and dynamic responses can be observed as shown in the previous cases when the droop coefficient is small. However, with the increase of the droop coefficient, the oscillating behavior would appear in the reference voltage amplitude and frequency, and similar dynamics can be observed in power sharing. The stability boundary with respect to the LBC delay time is much smaller, compared to the case when the LPF-based delay model is utilized, which infers that an optimistic conclusion might be deduced using the LPF-based delay model, whereas the lead-lag compensator-based delay model would result in conservative evaluation for the stability boundary.

8.5   Experimental Results

In order to validate the feasibility of the proposed enhanced hierarchical control strategy, the experimental results obtained from two parallel-connected DG units are presented and compared. The experimental setup consists of two 2.2 kW Danfoss inverters connected in parallel with linear and nonlinear loads, and dSPACE1106 platform was used to implement the control algorithms. The controller parameters of the MG are described in Table 8.2. The schematic of experimental setup is shown in Fig. 8.28, respectively.

The experimental results of the VSIs in the islanded MG system with and without using the proposed virtual impedance loops and AD method under resistive load conditions are compared in Figs. 8.29 and 8.30. As shown in Fig. 8.29, small output voltages of inverter 1 and inverter 2 would result in severe output currents oscillations, and higher output voltage reference would trip the converters due to the over-current protection. Figure 8.30 shows the experimental results of the MG under resistive load conditions, when the inner loop AD scheme with $k_{AD} = 28.5$ and the proposed virtual impedance loops with the virtual positive- and negative-sequence impedance, and the virtual variable harmonic impedance loops are used. It is observed that the oscillations of the output currents are alleviated with the proposed AD method and virtual impedance loops, which confirm the theoretical analysis.

Figure 8.31 shows the output currrent waveforms of the MG sharing the resistive load by using the droop method with the additional phase-shift loop (k_d) and AD (k_{AD}) method. Both inverters are sharing the resistive load in the normal operation mode, and a load step increase of 230 Ω is suddenly applied at $t = 3.35$ s and inverter 1 is disconnected at $t = 7.35$ s, while only inverter 2 is supplying the total load currents. As shown in Fig. 8.31, the inverter 2 can supply the load and the system remains stable when inverter 1 trips.

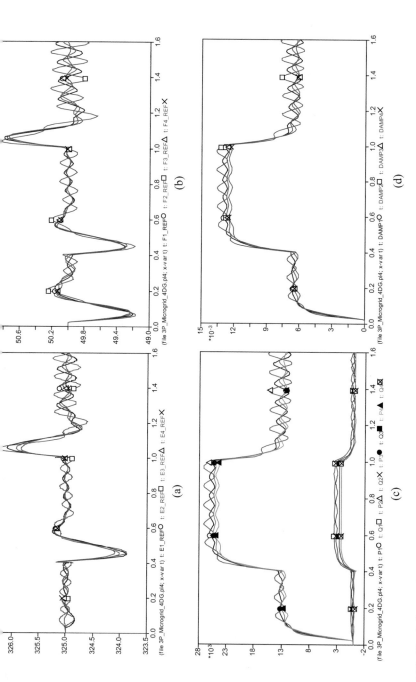

Fig. 8.26 Simulation results under nonidentical feeder inductance and resistance scenario with LBC delay $T_d = 50$ ms denoted using lead-lag compensator model when the drop coefficient $k_p = k_q = 0.0005$, secondary controller proportional gain $k_{pf} = k_{pe} = 0.1$ pu, secondary controller proportional gain $k_{pf} = k_{pe} = 0.1$ (the line inductances of DG2, DG3, DG4 are -0.1 pu, 0.1 pu, 0.2 pu perturbed from nominal values, and the line resistances of DG2, DG3, DG4 are 0.2 pu, 0.4 pu, 0.6 pu perturbed from nominal values). (**a**) Reference voltage amplitude of each DG; (**b**) reference frequency of each DG; (**c**) instantaneous active and reactive powers of each DG; (**d**) damping control output of each DG

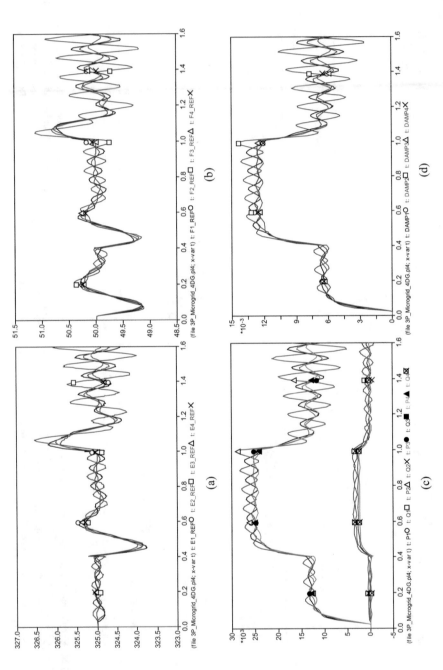

Fig. 8.27 Simulation results under nonidentical feeder inductance and resistance scenario with LBC delay $T_d = 60$ ms denoted using lead-lag compensator model when the droop coefficient $k_p = k_q = 0.0005$, secondary controller proportional gain $k_{pf} = k_{pe} = 0.1$ (the line inductances of DG2, DG3, DG4 are -0.1 pu, 0.1 pu, 0.2 pu perturbed from nominal values, and the line resistances of DG2, DG3, DG4 are 0.2 pu, 0.4 pu, 0.6 pu perturbed from nominal values). (**a**)

Table 8.2 System parameters of experimental setup

Symbol	Parameter	Value
v_{dc}, v_{MG}	DC and MG voltages	650 V, 311 V
f_0, f_s	MG and switching frequencies	50 Hz, 10 kHz
L, L_o	Filter and output inductances	1.8 mH, 1.8 mH
C	Filter capacitor	25 µF
R_L	Balanced resistive load	115/230 Ω
L_{NL}, R_{NL}, C_{NL}	Nonlinear load	84 µH, 235 µF, 460 Ω
k_{AD}	Active damping coefficient	28.5
$k_{pv}, k_{rv}, k_{5,7,11,13v}$	Voltage loop PR parameters	0.175, 200, 50, 40, 20, 20
$k_{pi}, k_{ri}, k_{5,7,11,13i}$	Current loop PR parameters	3, 50, 10, 5, 5

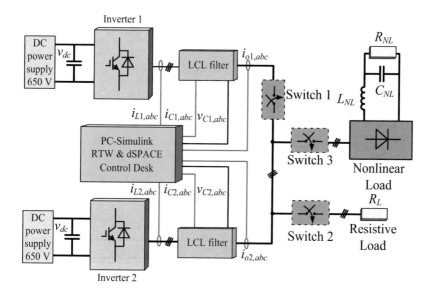

Fig. 8.28 Schematic of the experimental setup using parallel DG units

Figure 8.32 shows a comparison of the experimental results of the islanded MG system under nonlinear load conditions with and without using the virtual impedance loops which contain the virtual positive- and negative-sequence impedance, and the virtual variable harmonic impedance loops and harmonic compensation in the voltage and current loops. As shown in Fig. 8.32a, when the harmonic compensation is not activated and only the virtual positive- and negative-sequence impedance loops are activated, the output voltages are severely distorted by nonlinear loads and the total harmonic distortion (THD) of output voltage is about 5.45%. Figure 8.32b shows the experimental results when harmonic compensation is enabled and the proposed control method is used, where the multiple PR controllers are tuned at the 5th, 7th, 11th, and 13th harmonic frequencies in the voltage and current loops. In this case, THD of output voltage is reduced to 1.20%. It can be

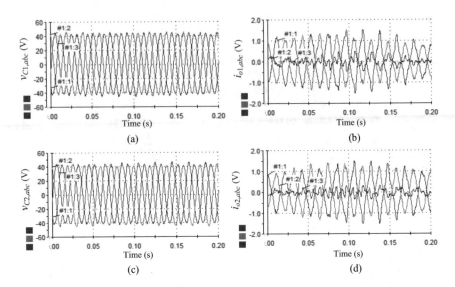

Fig. 8.29 Experimental results of the islanded MG system without using the proposed control method under resistive load conditions. (**a**) The output voltages of inverter 1. (**b**) The output currents of inverter 1. (**c**) The output voltages of inverter 2. (**d**) The output currents of inverter 2

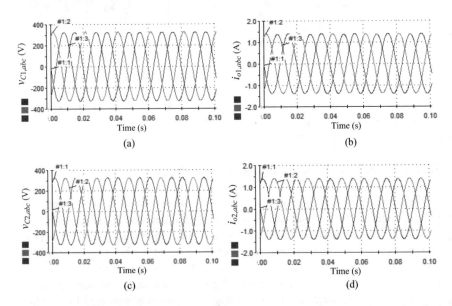

Fig. 8.30 Experimental results of the islanded MG system with using the proposed control method under resistive load conditions. (**a**) The output voltages of inverter 1. (**b**) The output currents of inverter 1. (**c**) The output voltages of inverter 2. (**d**) The output currents of inverter 2

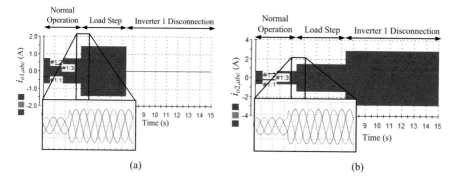

Fig. 8.31 Transient response of the output currents during resistive load step changes ($t = 3.53$ s) and sudden disconnection of inverter 1 ($t = 7.35$ s). (**a**) The output currents of inverter 1. (**b**) The output currents of inverter 2

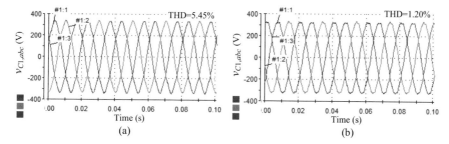

Fig. 8.32 Experimental results of the output voltages of a VSI. (**a**) The output voltages of inverter 1 without using the proposed method. (**b**) The output voltages of inverter 1 with using the proposed method

concluded that the THDs of the output voltages are effectively reduced with the proposed control strategies which contain AD method, harmonic compensation, and the virtual positive- and negative-sequence impedance, and the virtual variable harmonic impedance loops.

Figure 8.33 shows the performance of MG when the control of virtual positive- and negative-sequence impedance, and the virtual variable harmonic impedance loops are not activated. It is shown that the currents of inverter 1 and 2 are not identical, due to the circulating currents between these DG units. When the power sharing errors are compensated by using the proposed method, the transient response of the currents under nonlinear loads are given in Fig. 8.34. In Figs. 8.33 and 8.34, it is shown that the current sharing errors are effectively reduced and inverters 1 and 2 have similar output currents and the reactive and harmonic power sharing performances are improved. Note that the virtual harmonic impedance at the dominant harmonic frequencies, that is, 5th, 7th, 11th, and 13th are controlled. The higher harmonic frequencies can also be controlled in the virtual variable harmonic impedance loop when needed. The stability of the MG is also guaranteed under nonlinear

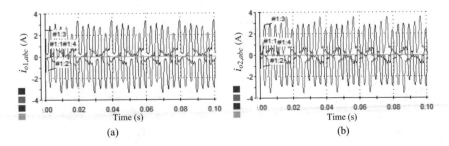

Fig. 8.33 Experimental results of the islanded MG system with conventional droop control. (**a**) The output currents of inverter 1. (**b**) The output currents of inverter 2

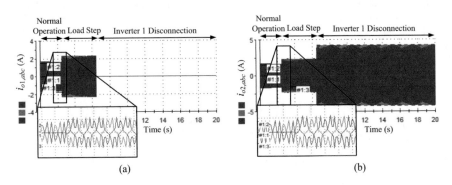

Fig. 8.34 Transient response of the output currents during load step changes ($t = 3.53$ s) and sudden disconnection of inverter 1 ($t = 7.35$ s) under nonlinear loads with the proposed virtual impedance loop and harmonic compensation controls. (**a**) The output currents of inverter 1. (**b**) The output currents of inverter 2

load conditions when the load step and sudden disconnection of one DG unit are applied.

The experimental results of the islanded MG system with and without using the secondary controller for the proposed control scheme under resistive load conditions are shown in Fig. 8.35. Figure 8.35a–c shows the experimental results without using the secondary controller and when a load change is suddenly applied at $t = 3.35$ s and the first inverter is disconnected at $t = 7.35$ s. As shown in Fig. 8.35a, the active power (P_1, P_2) and reactive power (Q_1, Q_2) sharings of the two DG units are achieved. A small amount of reactive power can be observed due to the effect of output inductance..

As depicted in Fig. 8.35b, the peak voltages of inverter 1 and 2 are not exactly the same under normal operation conditions, and small deviation is observed. With an increase of the load applied at $t = 3.35$ s, the peak voltage of inverter 2 drops and the voltage of inverter 1 increases. When inverter 1 is switched off at $t = 7.35$ s, the peak voltage of inverter 2 drops again due to an increase of load. As shown in Fig. 8.35c, the frequency is deviated from 50 Hz, that is, a steady-state error about 0.005 Hz can be observed under normal operation conditions. With an increase of

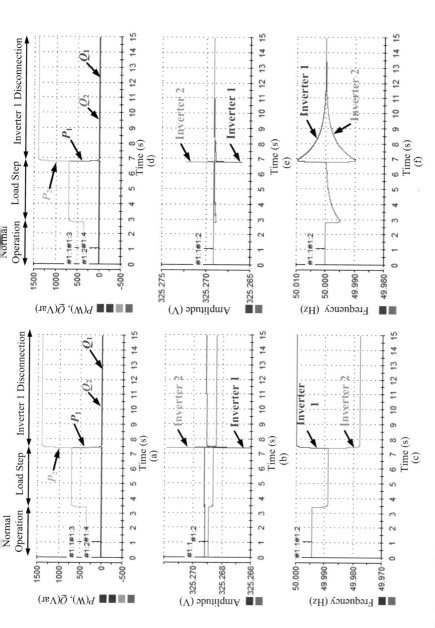

Fig. 8.35 Performance of the islanded MG consists of two DGs without (**a–c**) and with (**d**, **e**) using secondary controller under the resistive load conditions. (**a**, **d**) Active and reactive powers. (**b**, **e**) Voltage amplitude. (**c**, **f**) Frequency

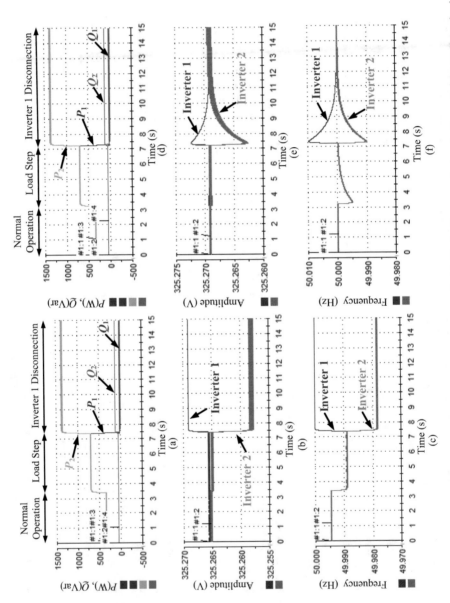

Fig. 8.36 Performance of the islanded MG consists of two DGs without (**a–c**) and with (**d, e**) using secondary controller under the nonlinear load conditions.

load, the frequencies of both inverters 1 and 2 drop for about 0.006 Hz. After the tripping of inverter 1, the frequency of the MG drops for 0.01 Hz to reach a steady state of 49.976 Hz. Figure 8.35d–f shows the performance of the islanded MG using the secondary controller. Figure 8.35d shows the dynamic response of the active power and reactive power in the DGs for each scenario. Note that there is a small increase in active power to restore the frequency deviation when the secondary control is activated.

The effect of the secondary control strategy to restore voltage and frequency deviations of the DGs is depicted in Fig. 8.35e, f. Notice that the deviations in voltage amplitude and frequency due to droop control and virtual impedance loops are recovered to the nominal values. Figure 8.35e shows the peak voltages of inverters 1 and 2 are identical under the normal operation scenario. The voltage amplitude restoration can be observed when a sudden increase of load is applied, and voltage amplitude recovers to the nominal value successfully even after disconnection of inverter 1. Figure 8.35f shows that the frequencies of inverters 1 and 2 are controlled to 50 Hz simultaneously under normal operation conditions. When the loads are suddenly increased, both frequency curves drop and gradually recover to 50 Hz. In the last scenario, the frequency restoration of inverter 2 is also achieved when the inverter 1 is switched off from the MG at $t = 6.8$ s, which recovers to the predefined frequency after a few seconds.

Figure 8.36 shows the experimental results of the islanded MG system for the proposed control scheme under nonlinear load conditions. Figure 8.36a–c shows the performance of the MG without using secondary controller and when a balanced resistive load is suddenly applied at $t = 3.35$ s and inverter 1 is disconnected at $t = 7.35$ s. Fig. 8.36d–f shows the performance of the islanded MG using the secondary controller, respectively.

Figure 8.36a, d shows that the active powers and reactive powers can be shared between DGs by means of droop control and enhanced virtual impedance loop, no matter with or without using the secondary control. These results illustrate that the P-ω droop control is sufficient to share the active power once the virtual impedance loops and inner AD scheme are adopted, since the frequency is a global variable in the MG system. The proposed secondary control is able to keep the reactive power shared between DG units under load variations. After disconnection of the inverter 1 from the MG system in the last scenario, inverter 2 feeds the load currents by injecting the doubled active power. By comparing Fig. 8.36b, c e, f, it can be observed that frequency and voltage amplitude restoration of the DG units can be achieved by means of the secondary control strategy.

8.6 Conclusions

This chapter presents an enhanced hierarchical control for three-phase parallel-connected VSI-based islanded MGs. The proposed method utilizes the primary control which is based on the virtual impedance loops with the virtual positive- and negative-sequence impedance loops at fundamental frequency, and the virtual variable harmonic impedance loop at harmonic frequencies, and droop control scheme with an additional phase-shift loop to enhance the sharing of reactive power and harmonic power between the DG units. The moving average filter-based sequence decomposition has been proposed to accurately extract the fundamental positive and negative sequences, and harmonic components for the virtual impedance loops. With the centralized controller of the secondary control, the voltage amplitude and frequency restoration are achieved. The developed small-signal model for the primary and secondary controls shows that an overdamped feature of the power loop is achieved to improve the whole MG system damping.

A multiloop control strategy with the inner-loop AD method for resistive and nonlinear load conditions is also presented. The capacitor currents of the LCL filter are used as feedback signals to actively damp the high frequency resonances while an outer voltage loop with output virtual impedance regulates the output voltage and ensure system stability over a wide range of operating conditions. The simulation results obtained from EMTP under nonidentical line impedance scenario are presented, and the effect of the low-bandwidth communication (LBC) delay is also simulated and compared with ideal scenario. Experimental results from a reduced-scale parallel-connected DG unit verified the effectiveness of the proposed control strategies, which can be easily applied for practical MG systems.

References

1. Mahmud, M. A., Hossain, M. J., Pota, H. R., & Oo, A. M. T. (2014). Robust nonlinear distributed controller design for active and reactive power sharing in islanded microgrids. *IEEE Transactions on Energy Coversion, 29*(4), 893–903.
2. Avelar, H. J., Parreira, W. A., Vieira, J. B., de Freitas, L. C. G., & Alves Coelho, E. A. (2012). A state equation model of a single-phase grid-connected inverter using a droop control scheme with extra phase shift control action. *IEEE Transactions on Industrial Electronics, 59*(3), 1527–1537.
3. Savaghebi, M., Jalilian, A., Vasquez, J. C., & Guerrero, J. M. (2012). Secondary control scheme for voltage unbalance compensation in an islanded droop-controlled microgrid. *IEEE Transactions on Smart Grid, 3*(2), 797–807.
4. Bidram, A., & Davoudi, A. (2012). Hierarchical structure of microgrids control system. *IEEE Transactions on Smart Grid, 3*(4), 1963–1976.
5. Rasheduzzaman, M., Mueller, J., & Kimball, J. (2014). An accurate small-signal model of inverter-dominated islanded microgrids using dq reference frame. *IEEE Transactions on Power Electronics, 2*(4), 1070–1080.
6. IEEE Recommended Practices and Requirements for Harmonic Control in Electrical Power System, IEEE Std. 519, 1992.
7. Xu, J., Xie, S., & Tang, T. (2014). Active damping-based control for grid-connected LCL-filtered inverter with injected grid current feedback only. *IEEE Transactions on Industrial Electronics, 61*(9), 4746–4758.
8. Lu, X., Guerrero, J. M., Teodorescu, R., Kerekes, T., Sun, K., & Huang, L. (2011). Control of parallel-connected bidirectional AC-DC converters in stationary frame for microgrid application. *Proc. IEEE ECCE, 5*, 4153–4160.

9. Savaghebi, M., Hashempour, M. M., & Guerrero, J. M. (2014). Hierarchical coordinated control of distributed generators and active power filters to enhance power quality of microgrids. *Proc. IEEE RTUCON*, 259–264.

10. Wu, D., Tang, F., Dragicevic, T., Vasquez, J. C., & Guerrero, J. M. (2014). Autonomous active power control for islanded ac microgrids with photovoltaic generation and energy storage system. *IEEE Transactions on Energy Conversion, 29*(4), 882–892.

11. Guerrero, J. M., Vasquez, J. C., Matas, J., Castilla, M., Vicua, L. G. D., & Castilla, M. (2011). Hierarchical control of droop-controlled AC and DC microgrids—A general approach toward standardization. *IEEE Transactions on Industrial Electronics, 58*(1), 158–172.

12. Guerrero, J. M., Loh, P., Chandorkar, M., & Lee, T. (2013). Advanced control architectures for intelligent microgrids—Part I: Decentralized and hierarchical control. *IEEE Transactions on Industrial Electronics, 60*(4), 1263–1270.

13. Vasquez, J., Guerrero, J. M., Savaghebi, M., Eloy-Garcia, J., & Teodorescu, R. (2013). Modeling, analysis, and design of stationary reference frame droop controlled parallel three-phase voltage source inverters. *IEEE Transactions on Industrial Electronics, 60*(4), 1271–1280.

14. Han, Y., Shen, P., Zhao, X., & Guerrero, J. M. (2017). Control strategies for islanded microgrid using enhanced hierarchical control structure with multiple current-loop damping schemes. *IEEE Transactions on Smart Grid, 8*(3), 1139–1153.

15. Vijay, A. S., Parth, N., Doolla, S., & Chandorkar, M. C. (2021). An adaptive virtual impedance control for improving power sharing among inverters in islanded AC microgrids. *IEEE Transactions on Smart Grid*. https://doi.org/10.1109/TSG.2021.3062391., Early Access.

16. Liu, J., Bevrani, H., & Ise, T. (2021). A design-oriented Q-V response modeling approach for grid-forming distributed generators considering different operation modes. *IEEE Journal on Emerging Selected Topics in Power Electronics*. https://doi.org/10.1109/JESTPE.2021.3057517., Early Access.

Index

© Springer Nature Switzerland AG 2022
Y. Han, *Modeling and Control of Power Electronic Converters for Microgrid Applications*, https://doi.org/10.1007/978-3-030-74513-4

Printed in the United States
by Baker & Taylor Publisher Services